SRA

Algebra Readiness

Practice

SRA

Columbus, OH

Author
Sharon Griffin
Professor of Education and
Adjunct Associate Professor of Psychology
Clark University
Worcester, Massachusetts

SRAonline.com

 SRA

Copyright © 2008 SRA/McGraw-Hill.

Printed in the United States of America.

Send all inquiries to this address:
SRA/McGraw-Hill
4400 Easton Commons
Columbus, OH 43219

ISBN: 978-0-07-614523-2
MHID: 0-07-614523-9

1 2 3 4 5 6 7 8 9 QPD 13 12 11 10 09 08 07

The McGraw·Hill Companies

Contents

Unit 1

Contents

Unit 2

Contents

Unit 3

Contents

Unit 4

Contents

Unit 5

Contents

Unit 6

Name _____ Date _____

Place Value and Addition: Lesson 1A

Practice

Write the greater number.

1. 2,361   2,163 _____

3. 3,742   3,762 _____

2. 1,385   1,359 _____

4. 7,042   7,024 _____

For each pair, write the place value where the numbers differ.

5. 6,243   6,043

_____

7. 8,315   9,315

_____

6. 7,214   7,274

_____

8. 2,046   2,346

_____

Fill in the blank with a digit to make each sentence true.

9. $6,145 < 6,1\_3$

_____

11. $3,429 > 3,\_31$

_____

10. $5,432 > 5\_57$

_____

12. $728 < \_40$

_____

Name _____ Date _____

Place Value and Addition: Lesson 1B

Practice

Write the greater number.

1. 1,823   1,832 _____

3. 7,145   7,154 _____

2. 6,827   6,817 _____

4. 5,341   5,431 _____

For each pair, write the place value where the numbers differ.

5. 8,142   8,162

_____

7. 5,245   5,345

_____

6. 6,187   9,187

_____

8. 4,216   4,219

_____

Fill in the blank with a digit to make each sentence true.

9. 3,267 < 3,__41

_____

10. 5,261 > 5,2__0

_____

Name _____ Date _____

Place Value and Addition: Lesson 2A

Practice

For each pair, write the place value where the numbers differ.

1. 625   635

_____

2. 4,034   5,034

_____

3. 3,215   3,915

_____

4. 826   820

_____

Write these numbers in order from least to greatest.

5. 643;   634;   642;   624;   613

_____

6. 384;   348;   483;   434;   388

_____

7. 673;   637;   763;   376;   676

_____

8. 4,216;   4,612;   4,618;   4,681;   4,482

_____

9. 1,035;   1,350;   1,503;   1,053;   1,353

_____

10. 8,634;   8,436;   8,346;   8,643;   8,364

_____

Name _____ Date _____

Place Value and Addition: Lesson 2B

Practice

For each pair, write the place value where the numbers differ.

1. 8,024   8,224

_____

3. 7,125   7,127

_____

2. 5,429   7,429

_____

4. 6,352   6,392

_____

Write these numbers in order from least to greatest.

5. 76; 67; 73; 63; 66

_____

6. 82; 208; 28; 288; 802

_____

7. 143; 134; 341; 314; 144

_____

8. 673; 376; 637; 367; 736

_____

9. 2,136; 2,316; 2,216; 2,326; 2,236

_____

10. 4,135; 4,105; 4,503; 4,351; 4,531

_____

Name _____ Date _____

Place Value and Addition: Lesson 3A

Practice

Find the sum. Write your answer.

1. 316
      + 522

6. 138
      + 831

2. 312
      + 564

7. 625
      + 312

3. 214
      + 604

8. 451
      + 328

4. 516
      + 220

9. 443
      + 216

5. 342
      + 518

10. 625
       + 204

Place Value and Addition: Lesson 3B

Practice

Find the sum. Write your answer.

1.   1237
     + 5814

6.   824
     + 369

2.   658
     + 820

7.   654
     + 2351

3.   4215
     + 3824

8.   8261
     + 1359

4.   4315
     + 3266

9.   4217
     + 1058

5.   743
     + 8615

10.   1625
     + 2047

Name _____ Date _____

Place Value and Addition: Lesson 4A

Practice

Find each sum. Write your answer. Write how many times you would make trades if you were using base-ten Blocks.

1. 256
     + 145

_____ trades

5. 518
     + 298

_____ trades

2. 342
     + 521

_____ trades

6. 358
     + 824

_____ trades

3. 321
     + 824

_____ trades

7. 634
     + 270

_____ trades

4. 424
     + 618

_____ trades

8. 268
     + 785

_____ trades

Name _____ Date _____

Place Value and Addition: Lesson 4B

Practice

Find each sum. Write your answer. Write how many times you would make trades if you were using Base-Ten Blocks.

1.   326
       + 547

_____ trades

2.   224
       + 176

_____ trades

3.   158
       + 375

_____ trades

4.   381
       + 456

_____ trades

5.   246
       + 172

_____ trades

6.   814
       + 176

_____ trades

7.   675
       + 204

_____ trades

8.   754
       + 787

_____ trades

Place Value and Subtraction: Lesson 1A

Practice

Use the digits given to make the largest number possible using each digit one time. Write your answer.

1. 2, 4, 5 _____

4. 2, 1, 6 _____

2. 3, 5, 4 _____

5. 7, 9, 2 _____

3. 1, 6, 3 _____

6. 3, 0, 8 _____

Use the digits given to make the smallest number possible using each digit one time. Write your answer.

7. 3, 2, 6 _____

10. 4, 3, 5 _____

8. 6, 2, 7 _____

11. 7, 1, 9 _____

9. 2, 1, 4 _____

12. 6, 3, 8 _____

Name _____ Date _____

Place Value and Subtraction: Lesson 1B

Practice

Use the digits given to make the largest number possible using each digit one time. Write your answer.

1. 6, 7, 1 _____

4. 0, 4, 2 _____

2. 3, 9, 2 _____

5. 6, 1, 8 _____

3. 2, 3, 5 _____

6. 5, 9, 3 _____

Use the digits given to make the smallest number possible using each digit one time. Write your answer.

7. 4, 5, 2 _____

10. 4, 2, 8 _____

8. 3, 6, 2 _____

11. 8, 2, 4 _____

9. 3, 7, 1 _____

12. 7, 3, 5 _____

Name _____ Date _____

Place Value and Subtraction: Lesson 2A

Practice

Choose the two numbers from each group that have the greatest difference. Write and solve that number sentence.

1. 326   541   822

_____

4. 456   134   721

_____

2. 302   264   591

_____

5. 312   545   748

_____

3. 143   624   250

_____

6. 240   653   420

_____

Choose the two numbers from each group that have the smallest difference. Write and solve that number sentence.

7. 241   365   132

_____

10. 164   358   571

_____

8. 326   453   271

_____

11. 580   640   510

_____

9. 674   252   417

_____

12. 670   320   840

_____

Name _____ Date _____

Place Value and Subtraction: Lesson 2B

Practice

Choose the two numbers from each group that have the greatest difference. Write and solve that number sentence.

1. 414   621   220

_____

4. 351   220   515

_____

2. 246   648   420

_____

5. 216   420   152

_____

3. 145   532   726

_____

6. 412   752   146

_____

Choose the two numbers from each group that have the smallest difference. Write and solve that number sentence.

7. 134   625   415

_____

10. 216   375   425

_____

8. 616   420   514

_____

11. 515   721   312

_____

9. 630   428   534

_____

12. 140   570   330

_____

Name _____ Date _____

Place Value and Subtraction: Lesson 3A

Practice

Find each difference. Write your answer. Use addition to check your answer.

1.
```
   421
 -   8
 ─────
```
Check:
```
     8
 +
 ─────
   421
```

2.
```
   618
 -  73
 ─────
```
Check:
```
 +
 ─────
   618
```

3.
```
   256
 - 175
 ─────
```
Check:
```
 +
 ─────
   256
```

4.
```
   734
 - 660
 ─────
```
Check:
```
 +
 ─────
   734
```

5.
```
   212
 - 146
 ─────
```
Check:
```
 +
 ─────
   212
```

6.
```
   565
 - 348
 ─────
```
Check:
```
 +
 ─────
   565
```

Place Value and Subtraction: Lesson 3B

Practice

Find each difference. Write your answer. Use addition to check your answer.

1. 624
 − 8
 ‾‾‾‾‾

Check: 8
        +
        ‾‾‾‾‾
         624

2. 152
 − 37
 ‾‾‾‾‾

Check:
        +
        ‾‾‾‾‾
         152

3. 537
 − 14
 ‾‾‾‾‾

Check:
        +
        ‾‾‾‾‾
         537

4. 843
 − 56
 ‾‾‾‾‾

Check:
        +
        ‾‾‾‾‾
         843

5. 715
 − 99
 ‾‾‾‾‾

Check:
        +
        ‾‾‾‾‾
         715

6. 653
 −142
 ‾‾‾‾‾

Check:
        +
        ‾‾‾‾‾
         653

7. 257
 −238
 ‾‾‾‾‾

Check:
        +
        ‾‾‾‾‾
         257

8. 645
 −327
 ‾‾‾‾‾

Check:
        +
        ‾‾‾‾‾
         645

Practice

Write the missing digits in each subtraction problem.

1.
```
   3 2 _
 − 2 _ 5
 ───────
   _ 1 3
```

5.
```
   4 3 _
 − _ 5 2
 ───────
   1 _ 4
```

2.
```
   _ 5 4
 − 3 _ 1
 ───────
   _ 8 3
```

6.
```
   6 _ 4
 − 1 7 _
 ───────
   _ 4 9
```

3.
```
   1 _ 6
 −   5 _
 ───────
   1 _ 8
```

7.
```
   _ 3 5
 − 1 _ 8
 ───────
   1 8 _
```

4.
```
   3 _ 4
 − 2 1 _
 ───────
   _ 6 2
```

8.
```
   5 _ _
 − 2 0 8
 ───────
   _ 7 3
```

Practice

Write the missing digits in each subtraction problem.

1.
```
   1 3 _
 -   _ 4
 ───────
   _ 1 3
```

6.
```
   _ 4 4
 - 3 _ 8
 ───────
   2 8 _
```

2.
```
   2 _ 2
 - _ 5 3
 ───────
     8 _
```

7.
```
   4 6 _
 - 3 _ 5
 ───────
   _ 1 3
```

3.
```
   4 _ 7
 - 1 5
 ───────
   _ 7 1
```

8.
```
   3 _ 9
 - 1 6 _
 ───────
   _ 1 1
```

4.
```
   3 4 _
 - 1 _ 3
 ───────
   _ 6 2
```

9.
```
   6 2 _
 - _ 4 5
 ───────
   1 _ 2
```

5.
```
   _ 2 4
 - 1 _ 1
 ───────
   1 6 _
```

10.
```
   4 5 _
 - 1 _ 6
 ───────
   _ 8 5
```

Name _____ Date _____

Multiplication: Lesson 1A

Practice

Write the multiplication sentence modeled by each array.

1.

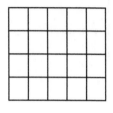

_____ × _____ = _____

4.

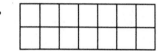

_____ × _____ = _____

2.

_____ × _____ = _____

5.

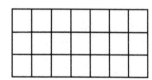

_____ × _____ = _____

3.

_____ × _____ = _____

6.

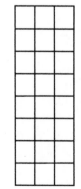

_____ × _____ = _____

Name _____ Date _____

Multiplication: Lesson 1B

Practice

Write the multiplication sentence modeled by each array.

1.

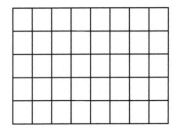

_____ × _____ = _____

4.

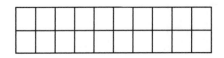

_____ × _____ = _____

2.

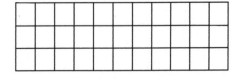

_____ × _____ = _____

5.

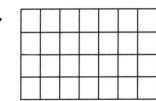

_____ × _____ = _____

3.

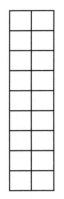

_____ × _____ = _____

6.

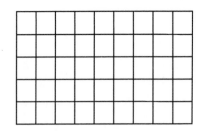

_____ × _____ = _____

Name _____ Date _____

Multiplication: Lesson 2A

Practice

Fill in the blanks to find each product.

1. 5×72

   (_____ × _____) + (_____ × _____)

   _____ + _____

   _____

3. 14×63

   (_____ × _____) + (_____ × _____)

   _____ + _____

   _____

2. 8×36

   (_____ × _____) + (_____ × _____)

   _____ + _____

   _____

4. 23×64

   (_____ × _____) + (_____ × _____)

   _____ + _____

   _____

Find each product. Write your answer.

5.   38
   $\times\ \ 5$

8.   742
   $\times\ \ 5$

6.   59
   $\times\ \ 6$

9.   47
   $\times\ \ 3$

7.   604
   $\times\ \ 9$

10.   82
   $\times\ 31$

Name _____ Date _____

Multiplication: Lesson 2B

Practice

Fill in the blanks to find each product.

1. 6 × 54

(_____ × _____) + (_____ × _____)

_____ + _____

_____

2. 4 × 67

(_____ × _____) + (_____ × _____)

_____ + _____

_____

3. 17 × 32

(_____ × _____) + (_____ × _____)

_____ + _____

_____

4. 34 × 72

(_____ × _____) + (_____ × _____)

_____ + _____

_____

Find each product. Write your answer.

5. 43
   × 9

6. 27
   × 4

7. 507
   × 6

8. 408
   × 7

9. 63
   × 8

10. 126
   × 34

Multiplication: Lesson 3A

Practice

Use the partial-product method to find each product. Show your work.
Write your answer.

1.
$$\begin{array}{r} 54 \\ \times\ 7 \\ \hline \end{array}$$

   + ___

2.
$$\begin{array}{r} 45 \\ \times\ 6 \\ \hline \end{array}$$

   + ___

3.
$$\begin{array}{r} 426 \\ \times\ 3 \\ \hline \end{array}$$

   + ___

4.
$$\begin{array}{r} 342 \\ \times\ 8 \\ \hline \end{array}$$

   + ___

5.
$$\begin{array}{r} 245 \\ \times\ 36 \\ \hline \end{array}$$

   + ___

6.
$$\begin{array}{r} 245 \\ \times\ 42 \\ \hline \end{array}$$

   + ___

Name _____ Date _____

Multiplication: Lesson 3B

Practice

Use the partial-product method to find each product. Show your work.
Write your answer.

1.   37
  × 4

  +____

2.   59
  × 4

  +____

3.   519
  × 6

  +____

4.   615
  × 7

  +____

5.   321
  × 27

  +____

6.   314
  × 53

  +____

Name _____ Date _____

Multiplication: Lesson 4A

Practice

Solve each multiplication problem. Write your answer. Then match the problem with the description of a grid that is large enough to hold the problem's array with the fewest number of squares left over. Use graph paper if you need help visualizing the grids.

1. $3 \times 8 =$ _____

A. a grid with 8 columns and 22 rows

2. $12 \times 6 =$ _____

B. a grid with 18 columns and 8 rows

3. $7 \times 21 =$ _____

C. a grid with 4 columns and 10 rows

4. $16 \times 5 =$ _____

D. a grid with 24 columns and 10 rows

5. $6 \times 18 =$ _____

E. a grid with 14 columns and 7 rows

6. $23 \times 8 =$ _____

F. a grid with 16 columns and 6 rows

Multiplication: Lesson 4B

Practice

Solve each multiplication problem. Write your answer. Then match the problem with the description of a grid that is large enough to hold the problem's array with the fewest number of squares left over. Use graph paper if you need help visualizing the grids.

1. $5 \times 12 =$ _____

   A. a grid with 8 columns and 5 rows

2. $7 \times 3 =$ _____

   B. a grid with 4 columns and 18 rows

3. $6 \times 20 =$ _____

   C. a grid with 5 columns and 15 rows

4. $3 \times 16 =$ _____

   D. a grid with 18 columns and 7 rows

5. $22 \times 9 =$ _____

   E. a grid with 7 columns and 20 rows

6. $17 \times 6 =$ _____

   F. a grid with 24 columns and 10 rows

Name _____ Date _____

Division: Lesson 1A

Practice

Solve each problem below using counters or drawing circles. Write a division number sentence to describe each situation.

1. Kacy is setting up the school dining room for a special luncheon. There are 6 tables in the room. If 42 people are coming to the luncheon, how many places should Kacy set at each table so the same number of guests are seated at each table?

   _____ guests at each table   _____

2. The sixth graders are organizing into 4 teams to play soccer. If there are 48 sixth graders, how many will be on each team?

   _____ sixth graders on each team   _____

3. One hundred fifty-five students are going on a field trip to the art museum. Five buses will be used to transport the students. How many students will be on each bus?

   _____ students will ride on each bus.   _____

4. Juanita is planting 60 tulip bulbs in 4 flower beds in her backyard. She wants to put the same number of bulbs in each bed. How many bulbs should be planted in each flower bed?

   _____ bulbs in each bed   _____

Division: Lesson 1B

Practice

Solve each problem below using Counters or drawing circles. Write a division number sentence to model each situation.

1. Forty students are organized into 8 teams for basketball drills. How many students will be on each team?

   _____ students on each team   _____

2. In the last 3 days, the high temperature increased a total of 18 degrees. The increase in temperature was the same for each day. How many degrees did the temperature increase each day?

   _____ degrees each day   _____

3. Forty-eight books need to be mailed in 6 cartons. The same number of books needs to be placed in each carton. How many books should be put into each box?

   _____ books in each carton   _____

4. José jogged a total of 28 miles in 4 days. Since he was training, he jogged the same number of miles each day. How many miles did he jog each day?

   _____ miles each day   _____

5. Sixteen chemistry students are put on 4 teams to work on a science fair project. How many students will be on each team?

   _____ students on each team   _____

6. Aretha has 12 thank-you notes to write for birthday presents she received. She wants to write the same number of notes each day for the next 4 days. How many notes should she write each day?

   _____ notes each day   _____

Name _____ Date _____

Division: Lesson 2A

Practice

Solve each problem below using Counters or drawing circles. Write a division number sentence to model each situation.

1. Flu vaccine comes in 30 cubic-centimeter vials. Each dose is 4 cubic centimeters. How many doses can be given from each vial?

_____

_____

2. Eldora has 20 daisy plants to put in the 3 flower beds in front of her house. She wants each bed to be the same, and she'll put any extra plants in her backyard. How many plants should she put in each bed in the front yard?

_____

_____

3. There are 23 fifth graders in Mr. Ong's physical education class. He organizes the students into 4 even teams. The remaining students act as referees. How many students will be on each team?

_____

_____

4. The hospitality committee at the community center is making fall gift baskets for 7 people. The committee has 30 apples to put in the baskets. How many apples should be put in each basket?

_____

_____

Name _____ Date _____

Division: Lesson 2B

Practice

Solve each problem below using Counters or drawing circles. Write a division number sentence to model each situation.

1. In preparation for the next 10-kilometer race, Marty wants to run a total of 26 miles this week. He will run the same distance on Monday, Wednesday, and Thursday, and finish the remainder of the miles on Friday. How many miles should he run on each of the first three days?

   _____

   _____

2. There are 5 display rooms in the local art gallery. The new exhibit has a total of 48 pieces of art. The curator will display the same number of pieces of art in each display room, and display the remainder of the pieces in the lobby. How many pieces of art will be in each display room?

   _____

   _____

3. The fourth graders are getting ready for a softball game. There are 27 students in the fourth grade class and they will be organized evenly into 2 teams. The remaining student will be the third-base coach. How many students will be on each team?

   _____

   _____

4. The school cafeteria staff is making up box lunches for a fifth grade picnic. They have 50 cookies to put into the box lunches. Each box is to get 3 cookies. How many boxes can they fill with the cookies?

   _____

   _____

Name _____ Date _____

Division: Lesson 3A

Practice

Solve each problem. Show your work. Write your answer.

1. Ms. Sanchez's students are coming to her classroom for the first day. The classroom has 8 rows of desks with 4 desks in each row. There are 30 students enrolled in her class. How many full rows of students will there be? How many students will sit in the last row?

_____ full rows of students with _____ student(s) in the last row

2. Mr. Perez is fixing fruit for his 3 children to take in their school lunches. He has 8 tangerines. How many tangerines can he put in each child's lunch? How many tangerines will he have left?

_____ tangerines in each lunch with _____ tangerine(s) left

3. Marlena is making costumes for her little nieces and nephews for trick-or-treat night. She has 11 yards of shiny silver cotton fabric, and each costume requires 2 yards of fabric. How many costumes can she make? How many yards of fabric will she have left?

_____ costumes with _____ yard(s) of fabric left

4. Tommy, a worker in the produce department at the grocery store, is putting together bunches of green onions. He'll put 8 onions in each bunch. The delivery this morning includes 260 green onions. How many bunches can he assemble? How many green onions will be left?

_____ bunches of green onions with _____ green onion(s) left

Name _____ Date _____

Division: Lesson 3B

Practice

Solve each problem. Show your work. Write your answers.

1. Tammy is organizing her CD library. She has 98 CDs to put on shelves that will each hold 8 CDs. How many full shelves will she have? How many CDs will she have left?

   _____ full shelves of CDs with _____ CD(s) left

2. Absha is making practice skirts for the students in her dance class. Each skirt requires 2 yards of fabric. She has 25 yards of pink net material to use to make the skirts. How many skirts can she make? How many yards of material will be left?

   _____ practice skirts with _____ yard(s) left

3. The service club at the local high school is gathering canned goods to send to victims of the recent earthquake. Each carton will hold 8 large cans of juice. How many cartons will they need to pack to hold the 75 large cans of juice they have collected? How many large cans of juice will be left?

   _____ cartons of large juice cans with _____ can(s) of juice left

4. Don's Roofing Service has received a shipment of 25 rolls of roofing paper to use for their next few jobs. Each job requires 4 rolls of roofing paper. How many jobs can they complete with this roofing paper? How many rolls will be left?

   _____ roofing jobs with _____ roll(s) left

5. In Tony's coin collection there are 118 aluminum cents. These cents can be put into plastic holders that will hold 6 cents each. How many plastic holders will he need for his groups of Indian Head cents? How many cents will be left?

   _____ plastic holders with _____ cent(s) left

6. Xander is putting his baseball cards into stacks of 8 cards each so that he can make trades with his friends. He has 55 cards that he wants to trade. How many stacks can he make? How many cards will be left?

   _____ stacks of cards with _____ card(s) left

Division: Lesson 4A

Practice

Find the remainder in each situation. Then determine whether it makes sense to round up, round down, or divide the remainder evenly. Explain. Write your answer.

1. To play a certain card game, you take the cards in the deck (there are 52 cards in a deck) and pass them out evenly to all 6 players. The cards that are left you place the facedown in the "discard" pile.

   _____

   _____

2. Each player will have _____ cards.

3. Carolyn uses 12 quilt squares to make a baby quilt. She has 50 quilt squares ready to use when the next baby is born.

   _____

   _____

   _____

4. Carolyn can make _____ baby quilts with the squares she has.

5. A local farmer has brought 6 dozen apples to the primary school to be used for snacks for the children. There are 4 classrooms in the primary school.

   _____

   _____

   _____

6. Each classroom can have _____ dozen apples.

Practice

Find the remainder in each situation. Then determine whether it makes sense to round up, round down, or divide the remainder evenly. Explain. Write your answer.

1. The band is taking school vans to this week's football game. Each van can hold 9 students. There are 98 students in the band.

   _____

   _____

   _____

2. The band will need _____ vans.

3. Clarissa uses 8 triangular pieces of cloth to make a placemat. She has 35 triangular pieces cut and ready to use.

   _____

   _____

   _____

4. Clarissa can make _____ placemats from the pieces she has.

5. Don has $63 in his change jar to give to his 6 grandchildren.

   _____

   _____

   _____

6. Each child will receive _____.

Name _____ Date _____

Parts of a Whole: Lesson 1A

Practice

Write each fraction modeled.

1.

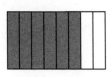

_____

2.

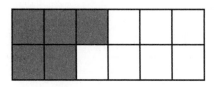

_____

3.

_____

4.

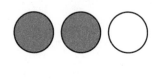

_____

Follow each instruction.

5. Divide the whole set below into 3 parts by circling individual sets.

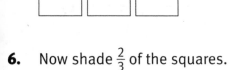

6. Now shade $\frac{2}{3}$ of the squares.

7. Below is $\frac{3}{4}$ of a set of pentagons. Draw the number of pentagons needed for a complete set.

8. Below is $\frac{1}{2}$ of a set of squares. Draw the number of squares needed for a complete set.

Name _____ Date _____

Parts of a Whole: Lesson 1B

Practice

Write each fraction shown.

1.

_____

2.
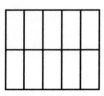

_____

3.

_____

4.

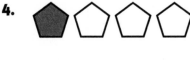

_____

Follow each instruction.

5. Divide the whole set below into 4 parts by circling individual sets.

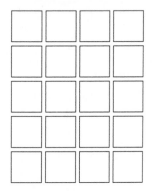

6. Now shade $\frac{3}{4}$ of the squares.

7. Below is $\frac{2}{5}$ of a set of circles. Draw the number of circles needed for a complete set.

8. Below is $\frac{2}{6}$ of a set of triangles. Draw the number of triangles needed for a complete set.

Name _____ Date _____

Parts of a Whole: Lesson 2A

Practice

Divide each shape into the number of equal parts indicated by the denominator. Then shade the unit fraction as indicated by the numerator. Write the fractional amount that is not shaded.

1. $\frac{3}{4}$

_____ is not shaded.

3. $\frac{3}{8}$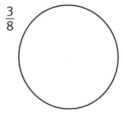

_____ is not shaded.

2. $\frac{4}{6}$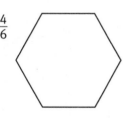

_____ is not shaded.

4. $\frac{7}{10}$

_____ is not shaded.

Draw each whole set.

5. Shown is $\frac{2}{7}$ of a set.

6. Shown is $\frac{3}{6}$ of a set.

Parts of a Whole: Lesson 2B

Practice

Divide each shape into the number of equal parts indicated by the denominator. Then shade the unit fraction as indicated by the numerator. Write the fractional amount that is not shaded.

1. $\frac{2}{5}$

_____ is not shaded.

3. $\frac{1}{6}$

_____ is not shaded.

2. $\frac{2}{3}$

_____ is not shaded.

4. $\frac{5}{8}$

_____ is not shaded.

Draw each whole set.

5. Shown is $\frac{2}{5}$ of a set.

6. Shown is $\frac{4}{9}$ of a set.

Name _____ Date _____

Parts of a Whole: Lesson 3A

Practice

Make one whole using the same unit fraction. Write how many more Fraction Bars you will need to make 1.

1.

1

$\frac{1}{4}$	$\frac{1}{4}$	$\frac{1}{4}$

_____ more $\frac{1}{4}$ Fraction Bars

2.

1

$\frac{1}{8}$	$\frac{1}{8}$	$\frac{1}{8}$

_____ more $\frac{1}{8}$ Fraction Bars

3.

1

$\frac{1}{6}$	$\frac{1}{6}$

_____ more $\frac{1}{6}$ Fraction Bars

Draw a grid for each of the following fractions.

4. $\frac{8}{8}$

5. $\frac{5}{5}$

Name _____ Date _____

Parts of a Whole: Lesson 3B

Practice

Make one whole using the same unit fraction. Write how many more Fraction Bars you will need to make 1.

1.

1

$\frac{1}{3}$	$\frac{1}{3}$

_____ more $\frac{1}{3}$ Fraction Bars

2.

1

$\frac{1}{10}$	$\frac{1}{10}$	$\frac{1}{10}$

_____ more $\frac{1}{10}$ Fraction Bars

3.

1

$\frac{1}{8}$	$\frac{1}{8}$	$\frac{1}{8}$	$\frac{1}{8}$	$\frac{1}{8}$	$\frac{1}{8}$

_____ more $\frac{1}{8}$ Fraction Bars

Draw a grid for each of the following fractions.

4. $\frac{4}{4}$

5. $\frac{12}{12}$

Name _____ Date _____

Parts of a Whole: Lesson 4A

Practice

Write each total using both cent notation and dollar notation.

1.

_____ _____

2.

_____ _____

3.

_____ _____

4.

_____ _____

5.

_____ _____

Name _____ Date _____

Parts of a Whole: Lesson 4B

Practice

Write each total using both cent notation and dollar notation.

1.

_____   _____

2.

_____   _____

3.

_____

4.

_____

5.

_____   _____

Name _____ Date _____

Positive and Negative Fractions: Lesson 1A

Practice

Do the fraction and the decimal match? Write *yes* or *no*.

1. $\frac{4}{5}$ and 0.5 _____

5. $\frac{4}{10}$ and 0.04 _____

2. $\frac{3}{8}$ and 0.375 _____

6. $\frac{9}{10}$ and 0.09 _____

3. $\frac{1}{4}$ and 0.25 _____

7. $\frac{15}{100}$ and 0.015 _____

4. $\frac{2}{3}$ and 0.6666 . . . _____

8. $\frac{7}{7}$ and 1.0 _____

Complete the table.

	Mixed Number	Improper Fraction	Decimal
9.	$1\frac{1}{5}$		
10.	$2\frac{2}{5}$		
11.	$1\frac{1}{3}$		
12.		$\frac{8}{5}$	
13.			2.75
14.			1.125
15.		$\frac{13}{4}$	
16.		$\frac{5}{3}$	

Name _____ Date _____

Positive and Negative Fractions: Lesson 1B

Practice

Do the fraction and the decimal match? Write *yes* or *no*.

1. $\frac{4}{8}$ and 0.5 _____

2. $\frac{1}{8}$ and 0.8 _____

3. $\frac{1}{3}$ and 0.33 _____

4. $\frac{7}{10}$ and 0.7777. . . _____

5. $\frac{7}{10}$ and 0.07 _____

6. $\frac{9}{10}$ and 0.9 _____

7. $\frac{21}{100}$ and 0.21 _____

8. $\frac{3}{3}$ and 1.0 _____

Complete the table.

	Mixed Number	Improper Fraction	Decimal
9.	$1\frac{2}{5}$		
10.	$2\frac{2}{3}$		
11.	$1\frac{1}{4}$		
12.		$\frac{9}{5}$	
13.			2.375
14.			1.5
15.		$\frac{11}{4}$	
16.		$\frac{11}{3}$	

Name _____ Date _____

Positive and Negative Fractions: Lesson 2A

Practice

Match each fraction with its decimal form. Draw a line from the fraction to the decimal.

1. $\frac{-3}{4}$ **A.** 3

2. $\frac{5}{8}$ **B.** −0.7

3. $\frac{17}{100}$ **C.** −0.625

4. $\frac{2}{5}$ **D.** −0.25

5. $\frac{-5}{8}$ **E.** −0.2

6. $\frac{-1}{5}$ **F.** 0.4

7. $\frac{-7}{10}$ **G.** −0.75

8. $\frac{3}{8}$ **H.** 0.375

9. $\frac{-1}{4}$ **I.** 0.17

10. $\frac{3}{1}$ **J.** 0.625

Name _____ Date _____

Positive and Negative Fractions: Lesson 2B

Practice

Match each fraction with its decimal form. Draw a line from the fraction to the decimal.

1. $\dfrac{-2}{3}$

           A. -0.625

2. $\dfrac{-3}{8}$

           B. -0.9

3. $\dfrac{-3}{5}$

           C. $-0.6666\ldots$

4. $\dfrac{5}{1}$

           D. 0.25

5. $\dfrac{-5}{8}$

           E. 0.375

6. $\dfrac{-23}{100}$

           F. -0.375

7. $\dfrac{-9}{10}$

           G. 0.625

8. $\dfrac{3}{8}$

           H. 5

9. $\dfrac{1}{4}$

           I. -0.6

10. $\dfrac{5}{8}$

           J. -0.23

Name _____ Date _____

Positive and Negative Fractions: Lesson 3A

Practice

Estimate where each fraction is on the number line. Label each with a point and the fraction.

1. $\frac{1}{4}$ and $\frac{2}{3}$

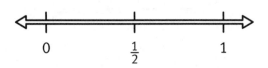

3. $\frac{4}{10}$ and $\frac{1}{3}$

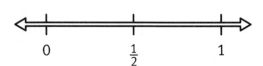

2. $\frac{2}{5}$ and $\frac{5}{8}$

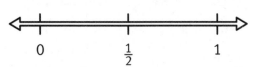

4. $\frac{5}{6}$ and $\frac{1}{8}$

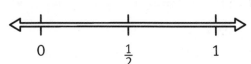

Name a decimal between the given decimals. Record it on the number line.

5.

1.5   2.5

6.

0.4   1.0

7.

0.25   0.5

8.

0.6   0.7

Name _____ Date _____

Positive and Negative Fractions: Lesson 3B

Practice

Estimate where each fraction is on the number line. Label each with a point and the fraction.

1. $\frac{1}{5}$ and $\frac{5}{8}$

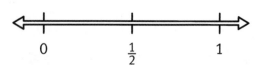

3. $\frac{9}{10}$ and $\frac{5}{12}$

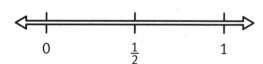

2. $\frac{1}{8}$ and $\frac{3}{4}$

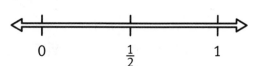

4. $\frac{3}{5}$ and $\frac{1}{3}$

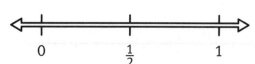

Name a decimal between the given decimals. Record it on the number line.

5.

6.

7.

8.

Name _____ Date _____

Positive and Negative Fractions: Lesson 4A

Practice

Estimate where each decimal is on the number line. Label each with a point and the decimal.

. 0.4 and 0.8

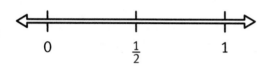

0     $\frac{1}{2}$ 1

3. −0.6 and −0.1

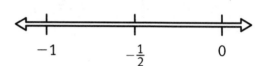

−1 $-\frac{1}{2}$ 0

. 0.125 and 0.75

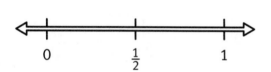

0     $\frac{1}{2}$ 1

4. −0.25 and −0.8

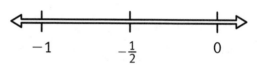

−1 $-\frac{1}{2}$ 0

Name a decimal between the given decimals. Record it on the number line.

.

0.4   0.8

5.

0.5   0.6

7.

−0.4   0.4

8.

−0.7   −0.65

Name _____ Date _____

Positive and Negative Fractions: Lesson 4B

Practice

Estimate where each decimal is on the number line. Label each with a point and the decimal.

1. 0.3 and 0.75

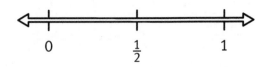

2. 0.4 and 0.9

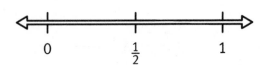

3. −0.6 and −0.4

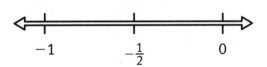

4. −0.1 and −0.75

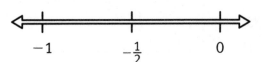

Name a decimal between the given decimals. Record it on the number line.

5.

0.35   0.6

6.

0.2   0.3

7.

−0.25   0.25

8.

−0.95   −0.8

Name _____ Date _____

Prime Factorization and Powers of 10: Lesson 1A

Practice

Write the factors in each expression.

. $3 \times 9 = 27$ _____ _____

3. $5 \times 12 = 60$ _____ _____

. $11 \times 6 = 66$ _____ _____

4. $8 \times 15 = 120$ _____ _____

Write the first five multiples of each number.

. 3 _____

7. 8 _____

. 12 _____

8. 7 _____

Write whether the following numbers are *prime* or *composite*.

. 10 is a _____ number.

0. 5 is a _____ number.

1. 17 is a _____ number.

2. 35 is a _____ number.

3. 27 is a _____ number.

4. 43 is a _____ number.

5. 9 is a _____ number.

Name _____ Date _____

Prime Factorization and Powers of 10: Lesson 1B

Practice

Write the factors in each expression.

1. $6 \times 4 = 24$ _____ _____

3. $8 \times 10 = 80$ _____ _____

2. $14 \times 3 = 42$ _____ _____

4. $7 \times 12 = 84$ _____ _____

Write the first five multiples of each number.

5. 6 _____

7. 11 _____

6. 9 _____

8. 15 _____

Write whether the following numbers are *prime* or *composite*.

9. 15 is a _____ number.

10. 7 is a _____ number.

11. 23 is a _____ number.

12. 6 is a _____ number.

13. 4 is a _____ number.

14. 41 is a _____ number.

15. 13 is a _____ number.

Chapter 7

Name _____ Date _____

Prime Factorization and Powers of 10: Lesson 2A

Practice

Write the following numbers as the product of their prime factors. Use counters, factor trees, mental math, or paper and pencil to determine the prime factors for each number.

1. 20 _____

2. 45 _____

3. 54 _____

4. 18 _____

5. 15 _____

6. 84 _____

7. 52 _____

8. 100 _____

9. 28 _____

10. 26 _____

Name _____ Date _____

Prime Factorization and Powers of 10: Lesson 2B

Practice

Write the following numbers as the product of their prime factors. Use counters, factor trees, mental math, or paper and pencil to determine the prime factors for each number.

1. 30 _____

2. 55 _____

3. 24 _____

4. 28 _____

5. 35 _____

6. 32 _____

7. 72 _____

8. 90 _____

9. 17 _____

10. 42 _____

Name _____ Date _____

Prime Factorization and Powers of 10: Lesson 3A

Practice

Write the following numbers as the product of their prime factors in expanded form and using exponents.

1. 8 = _____ = _____

2. 12 = _____ = _____

3. 54 = _____ = _____

4. 120 = _____ = _____

5. 35 = _____ = _____

6. 32 = _____ = _____

7. 72 = _____ = _____

8. 80 = _____ = _____

9. 44 = _____ = _____

10. 56 = _____ = _____

Name _____ Date _____

Prime Factorization and Powers of 10: Lesson 3B

Practice

Write the following numbers as the product of their prime factors in expanded form and using exponents.

1. 16 = _____ = _____

2. 24 = _____ = _____

3. 77 = _____ = _____

4. 80 = _____ = _____

5. 45 = _____ = _____

6. 100 = _____ = _____

7. 63 = _____ = _____

8. 52 = _____ = _____

9. 34 = _____ = _____

10. 150 = _____ = _____

Name _____ Date _____

Prime Factorization and Powers of 10: Lesson 4A

Practice

Write each number in expanded form, then in expanded form using powers of ten.

1. 452 = _____ =

_____

2. 2,684 = _____ =

_____

3. 0.37 = _____ =

_____

4. 0.683 = _____ =

_____

5. 49.35 = _____ =

_____

6. 724.26 = _____ =

_____

Write each number in standard form.

7. $7 \times 10^2 + 8 \times 10^1 + 4 \times 10^0 =$ _____

8. $6 \times 10^4 + 9 \times 10^2 + 3 \times 10^1 + 7 \times 10^0 =$ _____

9. $2 \times 10^0 + 5 \times 10^{-1} + 8 \times 10^{-2} + 4 \times 10^{-3} =$ _____

10. $8 \times 10^2 + 9 \times 10^0 + 6 \times 10^{-1} + 3 \times 10^{-2} =$ _____

Name _____ Date _____

Prime Factorization and Powers of 10: Lesson 4B

Practice

Write each number in expanded form, then in expanded form using powers of ten.

1. $873 =$ _____ $=$

_____

2. $5,104 =$ _____ $=$

_____

3. $0.692 =$ _____ $=$

_____

4. $7.24 =$ _____ $=$

_____

5. $189.56 =$ _____ $=$

_____

6. $302.48 =$ _____ $=$

_____

Write each number in standard form.

7. $7 \times 10^3 + 8 \times 10^2 + 9 \times 10^1 + 4 \times 10^0 =$ _____

8. $5 \times 10^4 + 3 \times 10^3 + 4 \times 10^1 + 5 \times 10^0 =$ _____

9. $4 \times 10^1 + 7 \times 10^0 + 3 \times 10^{-1} + 8 \times 10^{-2} + 1 \times 10^{-3} =$ _____

10. $8 \times 10^{-1} + 3 \times 10^{-2} + 4 \times 10^{-3} + 7 \times 10^{-4} =$ _____

Name _____ Date _____

Order Fractions: Lesson 1A

Practice

Write the numerator or the denominator to make equivalent fractions.

1. $\dfrac{2}{3} = \dfrac{4}{\square} = \dfrac{\square}{9} = \dfrac{\square}{\square}$

2. $\dfrac{5}{8} = \dfrac{10}{\square} = \dfrac{\square}{24} = \dfrac{\square}{\square}$

Compare the shaded part of each of these fractions, and write a true statement using $<$, $>$, or $=$.

3.
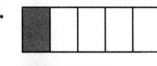

_____ _____ _____

4.
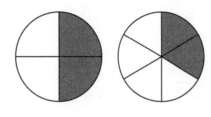

_____ _____ _____

Rewrite the fractions in each pair so they have the same denominator, and write a true statement using $<$, $>$, or $=$.

5. $\qquad \dfrac{1}{4} \qquad\qquad\qquad \dfrac{1}{3}$

_____ _____ _____

6. $\qquad \dfrac{5}{12} \qquad\qquad\qquad \dfrac{3}{10}$

_____ _____ _____

Compare the fractions using cross multiplication, and write a true statement using $<$, $>$, or $=$.

7. $\qquad \dfrac{6}{12} \qquad\qquad\qquad \dfrac{2}{4}$

$6 \times$ _____ ? $12 \times$ _____

_____ _____ _____

8. $\qquad \dfrac{3}{5} \qquad\qquad\qquad \dfrac{3}{7}$

$3 \times$ _____ ? $5 \times$ _____

_____ _____ _____

9. $\qquad \dfrac{1}{8} \qquad\qquad\qquad \dfrac{3}{11}$

$1 \times$ _____ ? $8 \times$ _____

_____ _____ _____

10. $\qquad \dfrac{3}{10} \qquad\qquad\qquad \dfrac{2}{6}$

$3 \times$ _____ ? $10 \times$ _____

_____ _____ _____

Order Fractions: Lesson 1B

Name _____ Date _____

Practice

Write the numerator or the denominator to make equivalent fractions.

1. $\dfrac{4}{5} = \dfrac{8}{\square} = \dfrac{\square}{15} = \dfrac{2}{\square}$

2. $\dfrac{3}{7} = \dfrac{6}{\square} = \dfrac{\square}{21} = \dfrac{\square}{\square}$

Compare the shaded part of each of these fractions, and write a true statement using $<$, $>$, or $=$.

3.
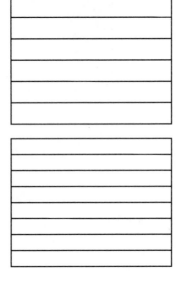

_____ _____ _____

4.
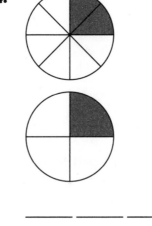

_____ _____ _____

Rewrite the fractions in each pair so they have the same denominator, and write a true statement using $<$, $>$, or $=$.

5. $\dfrac{2}{5}$　　　　$\dfrac{2}{3}$

_____ _____ _____

6. $\dfrac{5}{8}$　　　　$\dfrac{2}{5}$

_____ _____ _____

Compare the fractions using cross multiplication, and write a true statement using $<$, $>$, or $=$.

7. $\dfrac{2}{9}$　　　　　　$\dfrac{2}{7}$

$2 \times$ _____　　　? 　　$9 \times$ _____

_____ _____ _____

8. $\dfrac{4}{7}$　　　　　　$\dfrac{5}{9}$

$4 \times$ _____　　　? 　　$7 \times$ _____

_____ _____ _____

Name _____ Date _____

Order Fractions: Lesson 2A

Practice

Use mental math or play money to find the answer to each question.
Write your answers.

1. How many pennies are in $\frac{7}{10}$ of 100? _____

2. $\frac{7}{10} = \frac{\square}{100}$

3. What are the decimal equivalents? _____ _____

Write each as a part of 100 using fraction and decimal form.

4. 32 pennies

_____ _____

6. 65 pennies

_____ _____

5. 16 pennies

_____ _____

7. 83 pennies

_____ _____

Write each fraction as a decimal.

8. $\frac{3}{10}$

_____

12. $\frac{568}{1000}$

_____

9. $\frac{38}{100}$

_____

13. $\frac{19}{1000}$

_____

10. $\frac{47}{100}$

_____

14. $\frac{9}{100}$

_____

11. $\frac{9}{1}$

_____

15. $\frac{4}{5}$

_____

Name _____ Date _____

Order Fractions: Lesson 2B

Practice

Use mental math or play money to find the answer to each question. **Write your answers.**

1. How many pennies are in $\frac{3}{10}$ of 100? _____

2. $\frac{3}{10} = \frac{\square}{100}$

3. What are the decimal equivalents? _____ _____

Write each as a part of 100 using fraction and decimal form.

4. 36 pennies

_____ _____

6. 75 pennies

_____ _____

5. 52 pennies

_____ _____

7. 91 pennies

_____ _____

Write each fraction as a decimal.

8. $\frac{6}{10}$

_____

12. $\frac{17}{1000}$

_____

9. $\frac{4}{100}$

_____

13. $\frac{451}{1000}$

_____

10. $\frac{82}{100}$

_____

14. $\frac{66}{100}$

_____

11. $\frac{3}{1000}$

_____

15. $\frac{1}{4}$

_____

Name _____ Date _____

Order Fractions: Lesson 3A

Practice

Write the answer to each question. You can draw models to help you find percents.

20% of 35

1. 20% is the same as the fraction _____.

2. The fraction, _____, of 35 is _____.

3. 20% of 35 is _____.

25% of 40

4. 25% is the same as the fraction _____.

5. The fraction, _____, of 40 is _____.

6. 25% of 40 is _____.

10% of 80

7. 10% is the same as the decimal _____.

8. The decimal, _____, of 80 is _____.

9. 10% of 80 is _____.

Complete the chart below so the percent, fraction, and decimal in each row equal each other.

	Percent	Fraction	Decimal
10.	80%		0.8
11.	50%		0.5
12.	30%	$\frac{3}{10}$	
13.	5%		
14.	15%		

Name _____ Date _____

Order Fractions: Lesson 3B

Practice

Write the answer to each question. You can draw models to help you find percents.

10% of 40

1. 10% is the same as the fraction _____.

2. The fraction, _____, of 40 is _____.

3. 10% of 40 is _____.

50% of 88

4. 50% is the same as the fraction _____.

5. The fraction, _____, of 88 is _____.

6. 50% of 88 is _____.

30% of 60

7. 30% is the same as the decimal _____.

8. The decimal, _____, of 60 is _____.

9. 30% of 60 is _____.

Complete the chart below so the percent, fraction, and decimal in each row equal each other.

	Percent	Fraction	Decimal
10.	70%		0.7
11.	40%		0.4
12.	25%	$\frac{25}{100}$	
13.	9%		
14.	62%		

Order Fractions: Lesson 4A

Practice

Estimate where each number is on the number line. Label each with a point and the given value.

1. 0.3 and 0.75

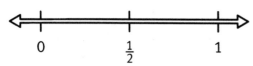

   0    $\frac{1}{2}$ 1

2. 20% and 65%

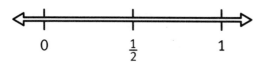

   0    $\frac{1}{2}$ 1

3. $\frac{2}{5}$ and $\frac{5}{8}$

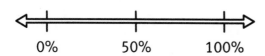

   0%   50%   100%

Name a percentage between the given values. Record it on the number line.

4.

   12%   45%

5.

   56%   66%

6.

   0.4   0.6

Name _____ Date _____

Order Fractions: Lesson 4B

Practice

Estimate where each number is on the number line. Label each with a point and the given value.

1. 0.45 and 0.8

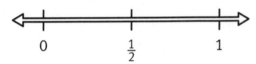

$$0 \qquad \frac{1}{2} \qquad 1$$

2. 30% and 85%

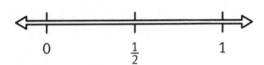

$$0 \qquad \frac{1}{2} \qquad 1$$

3. $\frac{1}{5}$ and $\frac{3}{4}$

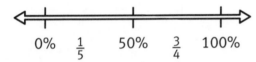

$$0\% \qquad \frac{1}{5} \qquad 50\% \qquad \frac{3}{4} \qquad 100\%$$

Name a percentage between the given values. Record it on the number line.

4.

$$35\% \qquad\qquad\qquad 60\%$$

5.

$$62\% \qquad\qquad\qquad 72\%$$

6.

$$0.44 \qquad\qquad\qquad 0.84$$

Name _____ Date _____

Adding and Subtracting Rational Numbers: Lesson 1A

Practice

Use the number line to help you solve each problem.

1. 12 − 4 = _____

2. −5 + −8 = _____

3. 5 + −2 = _____

4. −7 + 9 = _____

5. −3 − 6 = _____

6. 3 − −2 = _____

7. 8 − 14 = _____

8. 7 − −2 = _____

9. 7 + −3 = _____

10. −2 + −8 = _____

11. −6 − −5 = _____

12. 8 − 2 = _____

Name _____ Date _____

Adding and Subtracting Rational Numbers: Lesson 1B

Practice

Use the number line to help you solve each problem.

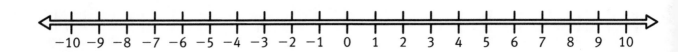

1. $10 - 3 =$ _____

2. $-6 + -2 =$ _____

3. $4 + -7 =$ _____

4. $-5 + 10 =$ _____

5. $-5 - 6 =$ _____

6. $5 + -3 =$ _____

7. $3 - 12 =$ _____

8. $12 - -4 =$ _____

9. $10 + -3 =$ _____

10. $-3 + -10 =$ _____

11. $-8 - -2 =$ _____

12. $-9 + 9 =$ _____

Name _____ Date _____

Adding and Subtracting Rational Numbers: Lesson 2A

Practice

Use grids to help you solve the addition and subtraction problems. Write your answers.

1. The basketball team has played $\frac{11}{20}$ of their scheduled games. What part of their scheduled games do they still have left to play?

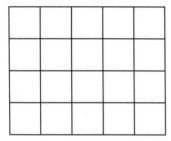

_____ of the games

2. It rained $\frac{1}{8}$ inch on Monday and $\frac{5}{8}$ inch on Tuesday. What was the total rainfall on these two days?

_____ inch

3. Tia has finished $\frac{3}{4}$ of her math homework problems. What part of her assignment does she still have to complete? Draw your own grid.

_____ of the assignment

4. Ruel has $\frac{3}{4}$ cup of milk in his measuring cup to make a frittata. The recipe says to use $\frac{1}{4}$ cup in the egg mixture and save the rest for the cheese layer. How much milk will be left in the measuring cup after Ruel makes the egg mixture? Draw your own grid.

_____ cup

Name _____ Date _____

Adding and Subtracting Rational Numbers: Lesson 2B

Practice

Use grids to help you solve the addition and subtraction problems. Write your answers.

1. Duan, the basketball team manager, is responsible for washing the team uniforms. There are 14 uniforms to be washed this week. He has already washed $\frac{9}{14}$ of the uniforms. What fraction of the uniforms does he still have to wash?

    _____ of the uniforms

2. It rained $\frac{7}{8}$ inch on Monday and $\frac{3}{8}$ inch on Tuesday. How much more did it rain on Monday than on Tuesday?

    _____ inch

3. The decorating committee finished $\frac{3}{10}$ of the table decorations for the upcoming festival on Tuesday and $\frac{4}{10}$ of the table decorations on Wednesday. What part of the decorations have they finished on these two days? Draw your grid.

    _____ of the decorations

4. Ling has $\frac{7}{8}$ quart of motor oil. He needs to use $\frac{5}{8}$ quart to mix with gasoline for his riding lawn mower. What part of a quart of oil will he still have left? Draw your own grid.

    _____ quart

Name _____ Date _____

Adding and Subtracting Rational Numbers: Lesson 3A

Practice

Use the grid to add or subtract the decimals.

1. 0.57 + 0.24 = _____

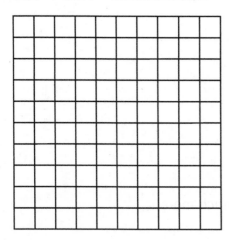

3. 0.63 − 0.45 = _____

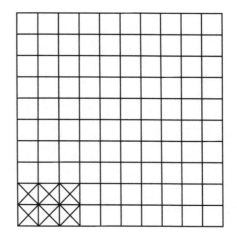

2. 0.88 − 0.26 = _____

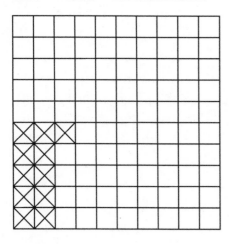

4. 0.71 − 0.39 = _____

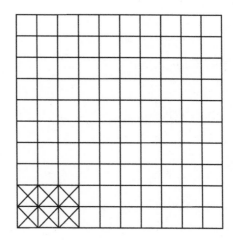

Add or subtract the decimals.

5. 0.36 + 0.08 = _____

6. −0.24 + −0.35 = _____

7. 0.57 − 0.26 = _____

8. 0.52 + −0.78 = _____

9. 0.49 + −0.16 = _____

10. −0.51 − −0.98 = _____

Adding and Subtracting Rational Numbers: Lesson 3B

Practice

Use the grid to add or subtract the decimals.

1. 0.68 + 0.13 = _____

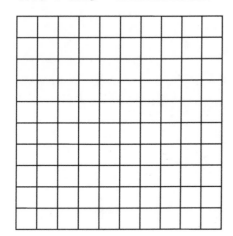

3. 0.34 − 0.19 = _____

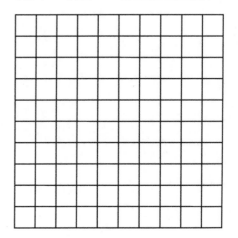

2. 0.74 − 0.27 = _____

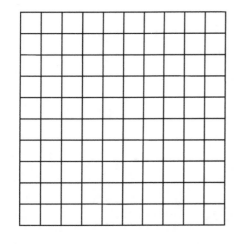

4. 0.52 − 0.08 = _____

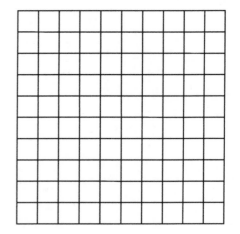

Add or subtract the decimals.

5. 0.63 + 0.28 = _____

6. −0.07 + −0.35 = _____

7. 0.49 − 0.33 = _____

8. 0.43 + −0.69 = _____

9. 0.42 + −0.79 = _____

10. −0.32 − −0.73 = _____

Name _____ Date _____

Adding and Subtracting Rational Numbers: Lesson 4A

Practice

Solve the equations.

1. $3^2 =$ _____

2. $8^0 =$ _____

3. $16^0 =$ _____

4. $10^3 =$ _____

5. $1^{23} =$ _____

6. $5^3 =$ _____

7. $27^1 =$ _____

8. $5^1 =$ _____

9. $1^2 =$ _____

10. $7^2 =$ _____

11. $4^3 =$ _____

12. $11^2 =$ _____

Name _____ Date _____

Adding and Subtracting Rational Numbers: Lesson 4B

Practice

Solve the equations.

1. $6^2 =$ _____

2. $4^0 =$ _____

3. $59^0 =$ _____

4. $10^4 =$ _____

5. $3^3 =$ _____

6. $13^2 =$ _____

7. $23^1 =$ _____

8. $17^1 =$ _____

9. $7^2 =$ _____

10. $1^4 =$ _____

11. $2^3 =$ _____

12. $7^0 =$ _____

Multiplying and Dividing Rational Numbers: Lesson 1A

Practice

Multiply or divide.

1. $-14 \times -5 = $ _____

2. $-63 \div -7 = $ _____

3. $45 \div -9 = $ _____

4. $42 \times 16 = $ _____

5. $28 \times -7 = $ _____

6. $-56 \div 8 = $ _____

7. $96 \div 12 = $ _____

8. $200 \div -40 = $ _____

9. $-8 \times 34 = $ _____

10. $-30 \times -12 = $ _____

Decide whether to multiply or divide in the following situations. Explain why you chose the operations you chose. Write your answers.

11. The stock for a company dropped 48 points over the last 6 days. On average, how many points did the stock drop each day?

-48 _____ $6 = $ _____

I chose this operation because _____

The stock dropped _____ points each day.

12. Sixteen students in the drama club have assembled programs for the school play. Each member assembled 25 programs. How many programs have they assembled altogether?

16 _____ $25 = $ _____

I chose this operation because _____

_____.

They assembled _____ programs.

Name _____ Date _____

Multiplying and Dividing Rational Numbers: Lesson 1B

Practice

Multiply or divide.

1. $45 \div -9 =$ _____

2. $100 \div 4 =$ _____

3. $-12 \times 8 =$ _____

4. $24 \times -2 =$ _____

5. $-36 \div -12 =$ _____

6. $35 \times 12 =$ _____

7. $-96 \div 3 =$ _____

8. $-72 \div -9 =$ _____

9. $-22 \times -5 =$ _____

10. $-42 \times -3 =$ _____

Decide whether to multiply or divide in the following situations. Explain why you chose the operations you chose. Write your answers.

11. Marsa practiced her clarinet an average of 35 minutes each day last week. How many minutes did she practice altogether last week?

    35 _____ $7 =$ _____

    I chose this operation because _____

    _____.

    She practiced _____ minutes last week.

12. The average low temperature dropped 30 degrees over the last 6 days. On average, how many degrees did the low temperature drop each day?

    -30 _____ $6 =$ _____

    I chose this operation because _____

    _____.

    The low temperature dropped an average of _____ degrees each day.

Multiplying and Dividing Rational Numbers: Lesson 2A

Practice

Reduce each fraction by dividing the numerator and the denominator by a common factor. Write your answer.

1. $\frac{12}{20} =$ _____

The common factor that I divided the denominator and numerator by was

_____.

2. $\frac{84}{72} =$ _____

The common factor that I divided the denominator and numerator by was

_____.

3. $\frac{35}{14} =$ _____

The common factor that I divided the denominator and numerator by was

_____.

4. $\frac{15}{42} =$ _____

The common factor that I divided the denominator and numerator by was

_____.

Multiply or divide the fractions. Write your answer as a reduced fraction.

5. $\frac{4}{3} \times \frac{9}{16} =$ _____

6. $\frac{2}{5} \div \frac{5}{6} =$ _____

7. $25 \div \frac{1}{5} =$ _____

8. $\frac{7}{10} \times \frac{15}{21} =$ _____

9. $\frac{5}{3} \div \frac{7}{9} =$ _____

10. $35 \times \frac{3}{7} =$ _____

11. $\frac{6}{7} \times \frac{21}{3} =$ _____

12. $\frac{10}{3} \div 6 =$ _____

Name _____ Date _____

Multiplying and Dividing Rational Numbers: Lesson 2B

Practice

Reduce each fraction by dividing the numerator and the denominator by a common factor. Write your answer.

1. $\dfrac{18}{30} =$ _____

The common factor that I divided the denominator and numerator by was

_____.

2. $\dfrac{32}{24} =$ _____

The common factor that I divided the denominator and numerator by was

_____.

3. $\dfrac{72}{12} =$ _____

The common factor that I divided the denominator and numerator by was

_____.

4. $\dfrac{45}{81} =$ _____

The common factor that I divided the denominator and numerator by was

_____.

Multiply or divide the fractions. Write your answer as a reduced fraction.

5. $\dfrac{3}{8} \times \dfrac{4}{15} =$ _____

6. $\dfrac{2}{5} \div \dfrac{5}{8} =$ _____

7. $20 \div \dfrac{1}{2} =$ _____

8. $\dfrac{6}{5} \times \dfrac{25}{18} =$ _____

9. $\dfrac{3}{4} \div \dfrac{12}{5} =$ _____

10. $20 \times \dfrac{3}{8} =$ _____

11. $\dfrac{2}{5} \times 10 =$ _____

12. $\dfrac{2}{3} \div 12 =$ _____

Name _____ Date _____

Multiplying and Dividing Rational Numbers: Lesson 3A

Practice

Use the grid to multiply the decimals.

1. 0.4 × 0.6 = _____

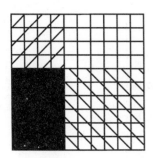

4. 0.3 × 0.9 = _____

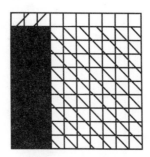

2. 0.5 × 0.5 = _____

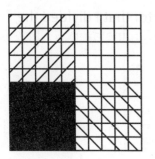

5. 0.8 × 0.1 = _____

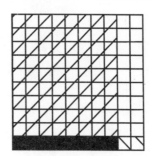

3. 0.2 × 0.7 = _____

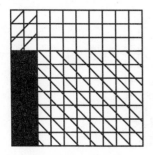

6. 0.7 × 0.6 = _____

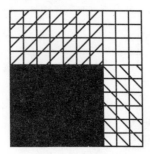

Multiply the decimals. Show your work, and write your answer.

7. Marta can jog 1 mile in 4.5 minutes. At this rate, how long will it take her to jog

6.2 miles? _____

8. A farmer can harvest 10.5 acres of wheat in 1 hour. How many acres can he harvest

in 8.5 hours? _____

Name _____ Date _____

Multiplying and Dividing Rational Numbers: Lesson 3B

Practice

Use the grid to multiply the decimals.

1. $0.5 \times 0.3 =$ _____

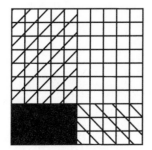

2. $0.6 \times 0.1 =$ _____

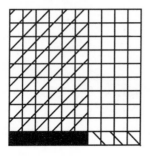

3. $0.2 \times 0.9 =$ _____

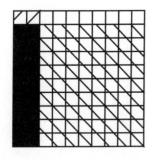

4. $0.3 \times 0.8 =$ _____

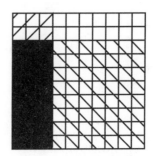

5. $0.7 \times 0.7 =$ _____

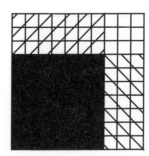

6. $0.9 \times 0.4 =$ _____

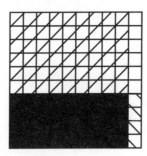

Multiply the decimals. Show your work, and write your answer.

7. An industrial printer can produce 12.5 labels every minute. How many labels can this machine produce in 45 minutes?

_____

8. A framing crew can frame a new house in 2.5 days. How many days will it take this crew to frame 7 new houses of the same design? _____

Name _____ Date _____

Multiplying and Dividing Rational Numbers: Lesson 4A

Practice

Multiply the decimals by the powers of ten. Write the product.

1. $100 \times 0.658 =$ _____

5. $1,000 \times 5.24 =$ _____

2. $10 \times 0.237 =$ _____

6. $100 \times 7.526 =$ _____

3. $10 \times 2.467 =$ _____

7. $0.01 \times 2.888 =$ _____

4. $10,000 \times 14.9 =$ _____

8. $0.1 \times 0.247 =$ _____

Divide the expressions as indicated. Write the result.

9. $0.63 \div 0.7 =$ _____

15. $0.144 \div 0.12 =$ _____

10. $4.2 \div 0.06 =$ _____

16. $1.32 \div 1.2 =$ _____

11. $45.45 \div 0.3 =$ _____

17. $560 \div 0.2 =$ _____

12. $36.36 \div 9 =$ _____

18. $0.0525 \div 0.25 =$ _____

13. $7.69 \div 0.769 =$ _____

19. $20 \div 0.8 =$ _____

14. $23.4 \div 117 =$ _____

20. $0.34 \div 0.2 =$ _____

Name _____ Date _____

Multiplying and Dividing Rational Numbers: Lesson 4B

Practice

Multiply the decimals by the powers of ten. Write the product.

1. $100 \times 6.728 = $ _____

2. $10 \times 16.52 = $ _____

3. $10 \times 0.089 = $ _____

4. $10{,}000 \times 9.43 = $ _____

5. $1{,}000 \times 2.458 = $ _____

6. $100 \times 0.0212 = $ _____

7. $0.1 \times 0.075 = $ _____

8. $0.01 \times 1.678 = $ _____

Divide the expressions as indicated. Write the result.

9. $0.56 \div 0.7 = $ _____

10. $8.1 \div 0.03 = $ _____

11. $16.16 \div 0.4 = $ _____

12. $55.66 \div 1.1 = $ _____

13. $14.7 \div -4.2 = $ _____

14. $42.23 \div 1.03 = $ _____

15. $0.324 \div 1.8 = $ _____

16. $1.32 \div 8 = $ _____

17. $425 \div 0.5 = $ _____

18. $0.0289 \div 0.17 = $ _____

19. $-308 \div -0.77 = $ _____

20. $129.5 \div -1.75 = $ _____

Name _____ Date _____

Understanding Rational Numbers: Lesson 1A

Practice

Complete the tables below so the percent, fraction, and decimal in each row equal each other.

	Percent	Fraction
1.	60%	
2.		$\frac{1}{4}$
3.	9%	
4.		$\frac{1}{2}$

	Percent	Decimal
5.	40%	
6.		0.18
7.	3%	
8.		0.55

	Fraction	Decimal
9.		0.17
10.	$\frac{3}{5}$	
11.	$\frac{9}{10}$	
12.		6.0

Understanding Rational Numbers: Lesson 1B

Practice

Complete the tables below so the percent, fraction, and decimal in each row equal each other.

	Percent	Fraction
1.	35%	
2.		$\frac{1}{2}$
3.	57%	
4.		$\frac{2}{5}$

	Percent	Decimal
5.	17%	
6.		0.02
7.	100%	
8.		0.32

	Fraction	Decimal
9.		0.17
10.	$\frac{3}{5}$	
11.	$\frac{5}{100}$	
12.		4.0

Understanding Rational Numbers: Lesson 2A

Practice

Convert the following decimals into fractions. Reduce the fractions to lowest terms. Write your answers.

. 0.08 = _____.

The reduced form is _____.

6. 1.25 = _____.

The reduced form is _____.

. 0.45 = _____.

The reduced form is _____.

7. 1.4 = _____.

The reduced form is _____.

. 0.42 = _____.

The reduced form is _____.

8. 0.148 = _____.

The reduced form is _____.

. 0.17 = _____.

The reduced form is _____.

9. 0.0025 = _____.

The reduced form is _____.

. 0.125 = _____.

The reduced form is _____.

10. 0.55 = _____.

The reduced form is _____.

Name _____ Date _____

Understanding Rational Numbers: Lesson 2B

Practice

Convert the following decimals into fractions. Reduce the fractions to lowest terms. Write your answers.

1. 0.44 = _____.

The reduced form is _____.

2. 0.35 = _____.

The reduced form is _____.

3. 0.39 = _____.

The reduced form is _____.

4. 0.23 = _____.

The reduced form is _____.

5. 0.175 = _____.

The reduced form is _____.

6. 1.75 = _____.

The reduced form is _____.

7. 1.2 = _____.

The reduced form is _____.

8. 0.225 = _____.

The reduced form is _____.

9. 0.005 = _____.

The reduced form is _____.

10. 1.15 = _____.

The reduced form is _____.

Understanding Rational Numbers: Lesson 3A

Practice

Follow the steps to solve the proportion.

There are 200 students in the middle school who are going on a field trip to the zoo. Fifteen percent of these students have family zoo passes. How many of the students have family zoo passes?

1. What is the unknown in the question?

_____

2. What fraction would you set up to represent the percent? _____

3. What fraction would you set up to represent the students who have passes (the part) divided by the total number of students going on the

trip (the whole)? _____

4. What is the proportional equation that you set up? How many

students have family zoo passes? _____

_____

Write a proportion for each problem. Write the solution to each problem.

5. There were 40 questions on the last history test. Diego got 85% of these questions right. How many questions did Diego answer correctly?

_____ _____

6. The basketball team scored 90 points in last night's game. Free throws accounted for 18 of those points. What percent of the points in last night's game came from free throws?

_____ _____

7. The drama club has assembled 150 programs for the upcoming play. This is 40% of all of the programs that need to be put together. How many programs do they need to assemble altogether?

_____ _____

8. Marquis has finished 36 of his math homework problems. There are 40 problems in the assignment. What percent of the problems has he finished?

_____ _____

Name _____ Date _____

Understanding Rational Numbers: Lesson 3B

Practice

Follow the steps to solve the proportion.

There are 160 students in the school who are going on a field trip to the aquarium. Twenty percent of these students have never been to the aquarium before. How many of the students have never been to the aquarium?

1. What is the unknown in the question?

   _____

2. What fraction would you set up to represent the percent? _____

3. What fraction would you set up to represent the students who have never been to the aquarium before (the part) divided by the total

   number of students going on the trip (the whole)? _____

4. What is the proportional equation that you set up? How many

   students have never been to the aquarium before? _____

   _____

Write a proportion for each problem. Write the solution to each problem.

5. There were 30 questions on the last math test. Dora got 90% of these questions right. How many questions did Dora answer correctly?

   _____   _____

7. This weekend's long distance bike trip is a 30 mile ride. Ling and Maio plan to ride 18 miles the first day. What percent of the trip do they plan to ride the first day?

   _____   _____

6. The football team scored 48 points in last Saturday's game. Field goals accounted for 12 of those points. What percent of the points in last Saturday's game came from free throws?

   _____   _____

8. Mindy's grade on her math quiz was 85%. She looked at her paper and realized that she had 17 problems correct. How many problems were on the quiz?

   _____   _____

Name _____ Date _____

Understanding Rational Numbers: Lesson 4A

Practice

Decide **whether the percent would make you pay more or less in the following situations, or whether the percent stands alone. Solve the problems. Write** *is added to the total, is subtracted from the total,* **or** *stands alone* **and the solution.**

. Manuella bought a pair of slacks at the store during a 30% off sale. If the slacks were originally 42 dollars, how much did she pay for them during the sale?

The percent _____. Therefore, the

amount she paid during the sale was _____.

. Thom and the salesman have agreed on a price for Thom's new car. The price they have agreed upon is 14,500 dollars. In addition, Thom will need to pay the sales tax, title fees, and dealer prep charges which are 12% of the agreed-upon price. How much will Thom pay altogether for the car?

The percent _____. Therefore, the

amount Thom will pay for the car is _____.

. Ms. Suhara owns shares of stock in a company that has done very well this year. She invested 6,500 dollars in this stock at the beginning of last year. The value of the stock increased 18% during the year. How much money did Ms. Suhara make on the stock last year?

The percent _____. Therefore, she

earned _____ on the stock last year.

. Given the information in Problem 3 above, what was the value of the stock at the beginning of this year?

The percent _____. Therefore, the value

of the stock at the beginning of this year was _____.

Name _____ Date _____

Understanding Rational Numbers: Lesson 4B

Practice

Decide whether the percent would make you pay more or less in the following situations, or whether the percent stands alone. Solve the problems. Write *is added to the total, is subtracted from the total,* or *stands alone* and the solution.

1. A furniture store offers a 10% discount to its customers who pay cash for their new furniture. The Hollis family is going to buy a new dining room table and chairs priced at 2,450 dollars. How much will they actually pay for the furniture since they have been saving for this purchase and can pay cash?

    The percent _____. The price they will

    pay for the furniture is _____.

2. Nick and Juanita went out for dinner to celebrate their anniversary. They received very good service and decided to tip 20%. The bill for their dinners originally was 68 dollars. How much will they pay altogether for this dinner?

    The percent _____. They will pay

    _____ for this dinner.

3. The students at the middle school are having a wrapping-paper sale to raise money for their school's new theater curtain. They will earn 30% of the value of the items that they sell. If the students sell 1,800 dollars during this fund-raising project, how much money will they raise to help to pay for the new curtain?

    The percent _____. Therefore, they will

    make _____ to help pay for the curtain.

4. The school supply store is having a 15% off sale on all of its top-of-the-line backpacks. How much will a backpack originally priced at 40 dollars sell for during this sale?

    The percent _____. Therefore, the price

    of the backpack during this sale will be _____.

Name _____ Date _____

Understanding Negative Exponents: Lesson 1A

Practice

Write the following exponential expressions as multiplication problems.
Simplify them by writing them in standard form.

1. 3^2

   In expanded form this is _____.

   Simplified into standard form this is

   _____.

2. 2^5

   In expanded form this is

   _____.

   Simplified into standard form this is _____.

3. 7^1

   In expanded form this is _____.

   Simplified into standard form this is

   _____.

4. 6^3

   In expanded form this is _____.

   Simplified into standard form this is

   _____.

5. 1^5

   In expanded form this is

   _____.

   Simplified into standard form this is

   _____.

6. 5^1

   In expanded form this is _____.

   Simplified into standard form this is

   _____.

7. 10^3

   In expanded form this is _____.

   Simplified into standard form this is

   _____.

8. 10^5

   In expanded form this is

   _____.

   Simplified into standard form this is

   _____.

9. 12^2

   In expanded form this is _____.

   Simplified into standard form this is

   _____.

10. 8^3

    In expanded form this is _____.

    Simplified into standard form this is

    _____.

Name _____ Date _____

Understanding Negative Exponents: Lesson 1B

Practice

Write the following exponential expressions as multiplication problems. Simplify them by writing them in standard form.

1. 5^2

In expanded form this is _____.

Simplified into standard form this is

_____.

2. 4^4

In expanded form this is

_____.

Simplified into standard form this is

_____.

3. 8^1

In expanded form this is _____.

Simplified into standard form this is

_____.

4. 1^3

In expanded form this is _____.

Simplified into standard form this is

_____.

5. 2^4

In expanded form this is _____.

Simplified into standard form this is

_____.

6. 15^1

In expanded form this is _____.

Simplified into standard form this is

_____.

7. 10^4

In expanded form this is

_____.

Simplified into standard form this is

_____.

8. 10^3

In expanded form this is _____.

Simplified into standard form this is

_____.

9. 7^3

In expanded form this is _____.

Simplified into standard form this is

_____.

10. 12^1

In expanded form this is _____.

Simplified into standard form this is

_____.

Chapter 12

Name _____ Date _____

Understanding Negative Exponents: Lesson 2A

Practice

Write the following exponential expressions in expanded form. Simplify them by writing them in standard form.

1. 3^{-2}

In expanded form this is _____.
Simplified into standard form this is

_____.

2. 4^{-4}

In expanded form this is _____.
Simplified into standard form this is

_____.

3. 8^{-1}

In expanded form this is _____.
Simplified into standard form this is
_____.

4. 1^{-5}

In expanded form this is

_____.

Simplified into standard form this is

_____.

5. 2^{-3}

In expanded form this is _____.
Simplified into standard form this is

_____.

6. 12^{-1}

In expanded form this is _____.
Simplified into standard form this is

_____.

7. 10^{-4}

In expanded form this is

_____.

Simplified into standard form this is

_____.

8. 10^{-3}

In expanded form this is _____.
Simplified into standard form this is

_____.

9. 5^{-2}

In expanded form this is _____.
Simplified into standard form this is

_____.

10. 9^{-2}

In expanded form this is _____.
Simplified into standard form this is

_____.

Understanding Negative Exponents: Lesson 2B

Practice

Write the following exponential expressions in expanded form. Simplify them by writing them in standard form.

1. 4^{-2}

In expanded form this is _____.
Simplified into standard form this is

_____.

2. 3^{-4}

In expanded form this is _____.
Simplified into standard form this is

_____.

3. 9^{-1}

In expanded form this is _____.
Simplified into standard form this is

_____.

4. 1^{-3}

In expanded form this is _____.
Simplified into standard form this is

_____.

5. 2^{-4}

In expanded form this is

_____.

Simplified into standard form this is

_____.

6. 15^{-1}

In expanded form this is _____.
Simplified into standard form this is

_____.

7. 10^{-3}

In expanded form this is _____.
Simplified into standard form this is

_____.

8. 10^{-5}

In expanded form this is

_____.

Simplified into standard form this is

_____.

9. 7^{-3}

In expanded form this is _____.
Simplified into standard form this is

_____.

10. 4^{-3}

In expanded form this is _____.
Simplified into standard form this is

_____.

Name _____ Date _____

Understanding Negative Exponents: Lesson 3A

Practice

Write the expression modeled by the fraction tiles using a negative exponent. Rewrite the expression as a fraction with a positive exponent.

1.

$\boxed{\dfrac{1}{8}} \times \boxed{\dfrac{1}{8}} \times \boxed{\dfrac{1}{8}}$

_____ = _____

2.

$\boxed{\dfrac{1}{6}} \times \boxed{\dfrac{1}{6}} \times \boxed{\dfrac{1}{6}} \times \boxed{\dfrac{1}{6}}$

_____ = _____

3.

$\boxed{\dfrac{1}{10}} \times \boxed{\dfrac{1}{10}} \times \boxed{\dfrac{1}{10}} \times \boxed{\dfrac{1}{10}} \times \boxed{\dfrac{1}{10}}$

_____ = _____

4.

$\boxed{\dfrac{1}{4}} \times \boxed{\dfrac{1}{4}} \times \boxed{\dfrac{1}{4}}$

_____ = _____

5.

$\boxed{\dfrac{1}{5}} \times \boxed{\dfrac{1}{5}} \times \boxed{\dfrac{1}{5}} \times \boxed{\dfrac{1}{5}}$

_____ = _____

6.

$\boxed{\dfrac{1}{2}} \times \boxed{\dfrac{1}{2}} \times \boxed{\dfrac{1}{2}} \times \boxed{\dfrac{1}{2}} \times \boxed{\dfrac{1}{2}}$

_____ = _____

Name _____ Date _____

Understanding Negative Exponents: Lesson 3B

Practice

Write the expression modeled by the fraction tiles using a negative exponent. Rewrite the expression as a fraction with a positive exponent.

1. $\boxed{\dfrac{1}{6}} \times \boxed{\dfrac{1}{6}}$

_____ = _____

2. $\boxed{\dfrac{1}{10}} \times \boxed{\dfrac{1}{10}} \times \boxed{\dfrac{1}{10}} \times \boxed{\dfrac{1}{10}}$

_____ = _____

3. $\boxed{\dfrac{1}{3}} \times \boxed{\dfrac{1}{3}} \times \boxed{\dfrac{1}{3}} \times \boxed{\dfrac{1}{3}} \times \boxed{\dfrac{1}{3}}$

_____ = _____

4. $\boxed{\dfrac{1}{4}} \times \boxed{\dfrac{1}{4}} \times \boxed{\dfrac{1}{4}} \times \boxed{\dfrac{1}{4}} \times \boxed{\dfrac{1}{4}} \times \boxed{\dfrac{1}{4}}$

_____ = _____

5. $\boxed{\dfrac{1}{2}} \times \boxed{\dfrac{1}{2}} \times \boxed{\dfrac{1}{2}} \times \boxed{\dfrac{1}{2}} \times \boxed{\dfrac{1}{2}} \times \boxed{\dfrac{1}{2}} \times \boxed{\dfrac{1}{2}}$

_____ = _____

6. $\boxed{\dfrac{1}{12}} \times \boxed{\dfrac{1}{12}} \times \boxed{\dfrac{1}{12}}$

_____ = _____

Name _____ Date _____

Understanding Negative Exponents: Lesson 4A

Practice

Multiply or divide the exponents. Write the result as an exponential expression.

1. $3^5 \times 3^2 =$ _____

2. $1^{23} \times 1^{15} =$ _____

3. $5^9 \div 5^3 =$ _____

4. $5^6 \times 5^7 =$ _____

5. $2^8 \div 2^8 =$ _____

6. $2^{-7} \times 2^5 =$ _____

7. $9^4 \div 9^3 =$ _____

8. $4^3 \times 4^7 =$ _____

9. $3^5 \div 3^7 =$ _____

10. $6^4 \div 6^7 =$ _____

11. $4^{10} \times 4^{10} =$ _____

12. $6^{-10} \div 6^4 =$ _____

Name _____ Date _____

Understanding Negative Exponents: Lesson 4B

Practice

Multiply or divide the exponents. Write the result as an exponential expression.

1. $4^5 \times 4^8 =$ _____

7. $4^4 \div 4^3 =$ _____

2. $7^3 \times 7^{15} =$ _____

8. $84 \times 84 =$ _____

3. $6^9 \div 6^3 =$ _____

9. $2^5 \div 2^9 =$ _____

4. $1^{15} \times 1^{25} =$ _____

10. $5^4 \div 5^4 =$ _____

5. $3^7 \div 3^9 =$ _____

11. $6^5 \times 6^5 =$ _____

6. $9^{-2} \times 9^{-3} =$ _____

12. $14^{-2} \div 14^{-2} =$ _____

Name _____ Date _____

Evaluating Expressions and Writing Equations: Lesson 1A

Practice

Draw parentheses around the operation that should be performed first in order to get the correct answer.

1. $6 + 12 \div 2 = 9$

4. $16 - 4 - 10 = 22$

2. $30 - 4 \div 2 = 28$

5. $20 \div 4 + 1 = 4$

3. $9 - 3 \times 2 = 3$

6. $36 \div 2 + 4 = 22$

Write the solution to each expression.

7. $40 - (8 - 12) =$ _____

12. $6 - (5 + 2) =$ _____

8. $(12 + 4) \div 4 =$ _____

13. $(2 - 9) + 13 =$ _____

9. $6 \times (7 + 3) =$ _____

14. $(8 + 2) - 10 =$ _____

10. $3 + (6 \times 5) =$ _____

15. $(2 - 7) + 5 =$ _____

11. $(5 \times 5) + 1 =$ _____

16. $1 - (8 + 9) =$ _____

Name _____ Date _____

Evaluating Expressions and Writing Equations: Lesson 1B

Practice

Draw parentheses around the operation that should be performed first in order to get the correct answer.

1. $5 + 8 \times 10 = 130$

4. $16 - 4 + 4 = 8$

2. $8 - (-4) - 5 = 17$

5. $4 \times 7 + (-2) = 20$

3. $12 - 5 - 15 = 22$

6. $12 + 8 \div 4 = 5$

Write the solution to each expression.

7. $(7 \times -3) - 4 = $ _____

12. $(14 - 4) + 8 = $ _____

8. $-3 + (-8 \times 2) = $ _____

13. $5 \times (10 + 5) = $ _____

9. $11 - (7 + 13) = $ _____

14. $(5 + 4) \times 3 = $ _____

10. $2 + (8 \times 9) = $ _____

15. $(9 - 7) + 8 = $ _____

11. $(9 \times 2) + 3 = $ _____

16. $2 - (6 + 8) = $ _____

Name _____ Date _____

Evaluating Expressions and Writing Equations: Lesson 2A

Practice

Use the digits to combine to reach the target number. Write each digit one time in the correct space.

1. Target number: 11

Digits to combine: 2, 3, 5, 6

Equation: _____

2. Target number: 2

Digits to combine: 2, 3, 4, −5

Equation: _____

3. Target number: 13

Digits to combine: 2, 3, 4, 7

Equation: _____

4. Target number: 1

Digits to combine: 2, 4, 5, 8

Equation: _____

5. Target number: −45

Digits to combine: −2, −3, 4, 5

Equation: _____

6. Target number: −10

Digits to combine: −2, 3, 5, 6

Equation: _____

Use the operations to reach the target number. Write each operation one time in the correct space.

7. Target number: 0

Operations to combine: + + −

Equation: (3 ___ 5) ___ (2 ___ 0)

8. Target number: 12

Operations to combine: + − ×

Equation: (6 ___ 4) ___ (5 ___ 2)

9. Target number: 200

Operations to combine: × × ÷

Equation: (20 ___ 2) ___ (5 ___ 4)

10. Target number: 1

Operations to combine: + ÷ ÷

Equation: (10 ___ 2) ___ (4 ___ 1)

Name _____ Date _____

Evaluating Expressions and Writing Equations: Lesson 2B

Practice

Use the digits to combine to reach the target number. Write each digit one time in the correct space.

1. Target number: −7

Digits to combine: 2, 3, 4, 5

Equation: _____

2. Target number: −5

Digits to combine: 1, 2, 4, 8

Equation: _____

3. Target number: 18

Digits to combine: 3, 0, 12, 2

Equation: _____

4. Target number: 3

Digits to combine: 9, 10, 11, 12

Equation: _____

5. Target number: −11

Digits to combine: −3, −1, 2, 6

Equation: _____

6. Target number: 8

Digits to combine: −3, −1, 2, 4

Equation: _____

Use the operations to reach the target number. Write each operation one time in the correct space.

7. Target number: −11

Operations to combine: + − ×

Equation: $(-8 \underline{\quad} 2) \underline{\quad} (9 \underline{\quad} 4)$

8. Target number: −4

Operations to combine: + − ÷

Equation: $(6 \underline{\quad} 2) \underline{\quad} (3 \underline{\quad} 5)$

9. Target number: −2

Operations to combine: + − −

Equation: $(-3 \underline{\quad} 2) \underline{\quad} (6 \underline{\quad} 3)$

10. Target number: −100

Operations to combine: × × ÷

Equation: $(-20 \underline{\quad} 4) \underline{\quad} (10 \underline{\quad} 2)$

Name _____ Date _____

Evaluating Expressions and Writing Equations: Lesson 3A

Practice

Evaluate the following expressions. Write the solution.

1. $4^2 \times (2 \times 10) + 7 =$ _____

4. $5^2 + 9 \times (8 - 8) =$ _____

2. $11 \times (5 - 3) + 4 =$ _____

5. $0 \times (7 + 11) + 4^2 =$ _____

3. $13 - (3^2 - 2) + 12 =$ _____

6. $6 + (3 - 9) \times 2^3 =$ _____

Draw a line from the equation to the correct answer.

7. $(4 + 2) - (3 \div 1) =$ **A.** 2

8. $(4 \div 2) \times (3 - 1) =$ **B.** 3

9. $4 - (2 - 3) \times 1 =$ **C.** 5

10. $(4 \times 2) \div (3 + 1) =$ **D.** 4

Evaluating Expressions and Writing Equations: Lesson 3B

Practice

Evaluate the following expressions. Write the solution.

1. $-5 + -3 \times (-4 - 2) =$ _____

2. $(14 - 9) \times 3 - 3 =$ _____

3. $6^2 - (4 \times 3) + 8 =$ _____

4. $5^2 \times 2 - (10 \times 5) =$ _____

5. $1^5 \times (4 \times -3) \div 2^2 =$ _____

6. $12 - (3 \times 7) + 1^2 =$ _____

Draw a line from the equation to the correct answer.

7. $(5 - 4) - (2 - 6) =$

A. 1

8. $(5 \times 4) \div 2 - 6 =$

B. 108

9. $5 + (4 \div 2) - 6 =$

C. 5

10. $(5 + 4) \times 2 \times 6 =$

D. 4

Evaluating Expressions and Writing Equations: Lesson 4A

Practice

Write the expressions and solutions that are described.

1. Luis is 4 years less than twice as old as his sister. If Louis is 12, then how old is his sister?

   _____

   His sister is _____ years old.

2. If you add 6 to an unknown number and multiply that sum by 7, you get 14. What is the number?

   _____

   The unknown number is _____.

3. If an unknown number is divided by 8 and then 7 is added to the quotient, the result is 10. What is the number?

   _____

   The unknown number is _____.

Write the solutions to the expressions that are described. It might be useful to use counters to model the expressions that are described.

4. The cost of a DVD is $4 more than three times the cost of a movie ticket. If the cost of the DVD is $17.50, then what is the price of the movie?

   _____

5. Marcella earned $16 less in tips than twice what Arvana earned. If Marcella earned $42, then how much did Arvana make in tips?

   _____

Name _____ Date _____

Evaluating Expressions and Writing Equations: Lesson 4B

Practice

Write the expressions and solutions that are described.

1. If you add 17 to an unknown number, you get one-third of 42. What is the number?

    _____

    The number is _____.

2. Angela's age is 3 years less than 4 times her sister's age. If Angela is 13, how old is her sister?

    _____

    Her sister is _____ years old.

3. If an unknown number is decreased by 10 then that difference is equal to −4. What is the number?

    _____

    The unknown number is _____.

Write the solutions to the expressions that are described. It might be useful to use counters to model the expressions that are described.

4. The cost of a football is $6 less than twice the cost of a soccerball. If the cost of the football is $18, how much does the soccerball cost?

    _____

5. If you multiply a number by 7 and then subtract −8, the result is −20?

    What is the number?   _____

Name _____ Date _____

Using Variables: Lesson 1A

Practice

Draw a line from the expression to the correct input/output.

1. $-b + 3$

   A. input: 2; output: -3

2. $5c + (-2)$

   B. input: 0; output: 3

3. $-2m + 1$

   C. input: 1; output: -3

4. $2p - 5$

   D. input: 1; output: 3

Write the expression that would give you the outputs desired.

5.

Input	Output
0	0
1	-2
2	-4
3	-6

Expression: _____

7.

Input	Output
0	2
1	3
2	4
3	5

Expression: _____

6.

Input	Output
0	0
1	1
2	8
3	27

Expression: _____

8.

Input	Output
0	-5
1	-4
2	-3
3	-2

Expression: _____

Name _____ Date _____

Using Variables: Lesson 1B

Practice

Draw a line from the expression to the correct input/output.

1. $-a + 4$

A. input: 2; output: 3

2. $5c - 8$

B. input: 0; output: -3

3. $-2n - 3$

C. input: 1; output: 3

4. $4p - 5$

D. input: 1; output: -3

Write the expression that would give you the outputs desired.

5.

Input	Output
0	4
1	5
2	6
3	7

Expression: _____

7.

Input	Output
0	0
1	-3
2	-6
3	-9

Expression: _____

6.

Input	Output
0	-2
1	-1
2	0
3	1

Expression: _____

8.

Input	Output
0	0
1	1
2	4
3	9

Expression: _____

Using Variables: Lesson 2A

Practice

Write the solution for the variable. It might be useful to use counters to model the situations.

1. $30 \div k = 6$

_____

4. $m - 9 = -2$

_____

2. $-3e = 15$

_____

5. $c - 4 = 10$

_____

3. $7 + j = -4$

_____

6. $7 + b = -3$

_____

Complete the table by writing the missing input or output for the expression $x - 6$.

	Input	Output
7.	-3	
8.		-6
9.		-2
10.	8	

Using Variables: Lesson 2B

Practice

Write the solution for the variable. It might be useful to use counters to model the situations.

1. $d - 7 = 12$

_____

4. $c \div 8 = 4$

_____

2. $9 + k = -1$

_____

5. $12 - e = 7$

_____

3. $-5m = 30$

_____

6. $-4y = 0$

_____

Complete the table by writing the missing input or output for the expression $x + 5$.

	Input	Output
7.	-7	
8.		3
9.	0	
10.		9

Name _____ Date _____

Using Variables: Lesson 3A

Practice

Draw a number line to represent the quantities described by the expressions.

1. 7 more than a number

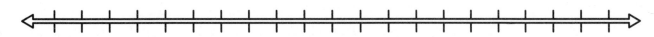

2. a number divided by 3

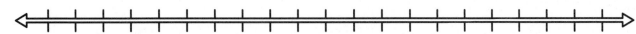

3. a number minus 4

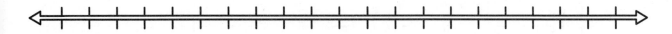

4. 3 times a number

Write the expression that is described.

5. 6 times a number

  _____

6. 5 more than a number

  _____ + _____ + _____

7. 8 divided by a number

  _____

8. a number minus 4

  _____

9. 10 less than a number

  _____

10. a number divided by 7

  _____

Name _____ Date _____

Using Variables: Lesson 3B

Practice

Draw a number line to represent the quantities described by the expressions.

1. 4 times a number

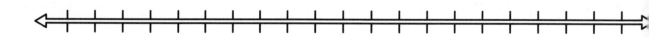

2. a number minus 8

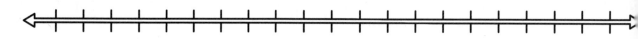

3. a number divided by 4

4. 6 more than a number

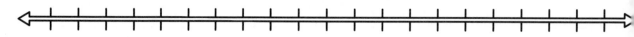

Write the expression that is described.

5. 6 more than a number

_____

6. 12 divided by a number

_____

7. 8 less than a number

_____

8. 4 times a number

_____

9. a number divided by −3

_____

10. 11 minus a number

_____

Name _____ Date _____

Using Variables: Lesson 4A

Practice

Draw a line from the description to the correct equation.

1. a number y is equal to the product of x and 9

2. a number y is equal to the sum of 9 and x

3. a number y is less than or equal to 9

4. a number y is equal to the difference of 9 and x

A. $y = 9 + x$

B. $y = 9 - x$

C. $y = 9x$

D. $y \le 9$

Write the equation that is described.

5. a number y is equal to x to the fourth power

_____

6. a number y is equal to the difference of 6 and a number x

_____

7. a number y is equal to the product of 6 and a number x

_____

8. a number y is greater than or equal to 10

_____

9. a number y is equal to the sum of 4 and x and also equal to the product of 6 and x

_____

10. a number y is greater than the product of 5 and x, and less than the sum of 5 and x

_____

Name _____ Date _____

Using Variables: Lesson 4B

Practice

Draw a line from the description to the correct equation.

1. a number y is equal to the quotient of x divided by 12

   A. $y = x - 12$

2. a number y is greater than 12

   B. $y = x \div 12$

3. a number y is equal to the product of 12 and a number x

   C. $y > 12$

4. a number y is 12 less than a number x

   D. $y = 12x$

Write the equation that is described.

5. a number y is less than -6

   _____

6. a number y is equal to the quotient of a number x divided by 3

   _____

7. a number y is greater than or equal to 8

   _____

8. a number y is equal to x to the third power

   _____

9. a number y is equal to the product of 5 and x and also equal to the sum of 5 and x

   _____

10. a number y is greater than the sum of 6 and x, and less than the product of 6 and x

   _____

Name _____ Date _____

Two-Variable Equations and Manipulating Symbols: Lesson 1A

Practice

Write the *x*-value in the table for the equation $y = 3x - 1$

x	y
	−10
	−7
	−4
	−1
	2
	5
	8

2. Write the *y*-value in the table for the equation $y = 4 + 3x$

x	y
−4	
−3	
−2	
−1	
0	
1	
2	

Create a table to represent some solutions for the following equations.

$y = x - 3$

x	y

5. $y = x + 2$

x	y

$x = 4$

x	y

6. $y = 2x - 1$

x	y

Name _____ Date _____

Two-Variable Equations and Manipulating Symbols: Lesson 1B

Practice

1. Write the *x*-value in the table for the equation $y = 4 - 2x$

x	y
	10
	8
	6
	4
	2
	0
	−2

2. Write the *y*-value in the table for the equation $y = 5x + 1$

x	y
−4	
−3	
−2	
−1	
0	
1	
2	

Create a table to represent some solutions for the following equations.

3. $y = x + 3$

x	y

5. $y = -3x - 1$

x	y

4. $y = -1$

x	y

6. $y = -x + 3$

x	y

Name _____ Date _____

Two-Variable Equations and Manipulating Symbols: Lesson 2A

Practice

Draw a line from the equations to the number that makes the statement true.

1. $a = b; a - 3 = b -$ _____   **A.** 5

2. $x = y; 1 + x = y +$ _____   **B.** -3

3. $j = k; j + 5 = k +$ _____   **C.** 1

4. $m = n; -3 - m =$ _____ $- n$ **D.** -5

5. $c = d; 0 + c = d +$ _____   **E.** 3

6. $u = v; u + -5 =$ _____ $+ v$ **F.** 0

Write the number that makes the statement true.

7. $p = q; p + 2 = q +$ _____   **10.** $m = n; m + 3 = n + (2 +$ _____$)$

8. $a = b; a - 4 = b +$ _____   **11.** $j = k; -3 - j =$ _____ $- k$

9. $r = s; r - 10 = s -$ _____   **12.** $c = d; 7 + c = d +$ _____

Name _____ Date _____

Two-Variable Equations and Manipulating Symbols: Lesson 2B

Practice

Draw a line from the equations to the number that makes the statement true.

1. $y = z; y + 4 = z +$ _____

    A. -5

2. $d = e; -5 + d =$ _____ $+ e$

    B. 2

3. $p = q; p + 0 =$ _____ $+ q$

    C. 6

4. $s = t; -s - 2 = -t -$ _____

    D. 4

5. $a = c; 6 + a = c +$ _____

    E. 3

6. $r = m; r - 3 = m -$ _____

    F. 0

Write the number that makes the statement true.

7. $a = c; a - 6 = c -$ _____

8. $r = t; r + 5 = t + (1 +$ _____ $)$

9. $d = g; d + 12 =$ _____ $+ g$

10. $b = j; -7 + b = j +$ _____

11. $h = k; h - 0 = k -$ _____

12. $m = w; m + (-2) = w -$ _____

Name _____ Date _____

Two-Variable Equations and Manipulating Symbols: Lesson 3A

Practice

aw a line from each equation to the number that makes the
atement true.

$y = z; -2 \times y = z \times$ _____

A. 4

$d = e; 4 \times d =$ _____ $\times e$

B. 2

$p = q; -3 \times p =$ _____ $\times q$

C. −1

$s = t; s \times 2 = t \times$ _____

D. −2

$a = c; -1 \times a = c \times$ _____

E. −4

$r = m; -4 \times r = m \times$ _____

F. −3

rite the number that makes the statement true.

$a = c; 5a =$ _____ c

10. $b = j; 0.3b = j \times$ _____

$r = t; -7r =$ _____ t

11. $h = k; h \div 4 =$ _____ k

$d = g; \frac{1}{5}d =$ _____ g

12. $m = w; 15m = ($ _____ $\times 3)w$

Name _____ Date _____

Two-Variable Equations and Manipulating Symbols: Lesson 3B

Practice

Draw a line from each equation to the number that makes the statement true.

1. $a = b; 3 \times a = b \times$ _____

A. 7

2. $x = y; x \times 7 = y \times$ _____

B. -2

3. $j = k; -3j = k \times$ _____

C. 6

4. $m = n; -2 \times m =$ _____ $\times n$

D. 3

5. $c = d; 6 \times c = d \times$ _____

E. 9

6. $u = v; u \times 9 =$ _____ $\times v$

F. -3

Write the number that makes the statement true.

7. $p = q; 4p =$ _____q

10. $m = n; 0.4m = m \times$ _____

8. $a = b; -2a =$ _____b

11. $j = k; j \div 2 =$ _____k

9. $r = s; \frac{1}{3}r =$ _____s

12. $c = d; 12c = ($ _____ $\times 3)d$

Name _____ Date _____

Two-Variable Equations and Manipulating Symbols: Lesson 4A

Practice

Write the operation that you selected to help you isolate the variable, and then perform it to both sides. Write the resulting equation.

1. $x - 5 = 10$

The operation I performed is

_____

The resulting equation is _____ = _____

2. $-3x = 12$

The operation I performed is

_____

The resulting equation is _____ = _____

3. $x \div 4 = 20$

The operation I performed is

_____

The resulting equation is _____ = _____

4. $12 + x = 15$

The operation I performed is

_____

The resulting equation is _____ = _____

Write the operations that you selected to help you isolate the variable, and then perform them to both sides. Write the resulting equations.

5. $4x - 7 = 13$

The first operation I performed is

_____

The resulting equation is _____ = _____

The next operation I performed is

_____

The resulting equation is _____ = _____

6. $-3x + 6 = -6$

The first operation I performed is

_____

The resulting equation is _____ = _____

The next operation I performed is

_____

The resulting equation is _____ = _____

Isolate the variable. Write the solution.

7. $x - 8 = 62$ _____ = _____

8. $x \div 4 = 12$ _____ = _____

9. $2x - 11 = 43$ _____ = _____

10. $-5x + 2 = 27$ _____ = _____

Two-Variable Equations and Manipulating Symbols: Lesson 4B

Practice

Write the operation that you selected to help you isolate the variable, and then perform it to both sides. Write the resulting equation.

1. $x + 4 = 11$

The operation I performed is

_____

The resulting equation is _____

2. $x \div 3 = -12$

The operation I performed is

_____

The resulting equation is _____

3. $-6x = -24$

The operation I performed is

_____

The resulting equation is _____

4. $-9 + x = 2$

The operation I performed is

_____

The resulting equation is _____

Write the operations that you selected to help you isolate the variable, and then perform them to both sides. Write the resulting equations.

5. $-2x + 11 = 17$

The first operation I performed is

_____

The resulting equation is _____

The next operation I performed is

_____

The resulting equation is _____

6. $6x - 4 = 38$

The first operation I performed is

_____

The resulting equation is _____

The next operation I performed is

_____

The resulting equation is _____

Isolate the variable. Write the solution.

7. $x + 10 = 34$ _____

8. $x \times 7 = -28$ _____

9. $3x - 1 = 23$ _____

10. $-2x + 6 = 0$ _____

Name _____ Date _____

Simplifying and Solving Equations: Lesson 1A

Practice

Apply the properties to simplify the expressions. Write the resulting expressions.

1. $(19 + 4) + 6$; associative property of addition

_____

2. $12 + (-12)$; inverse property of addition

_____

3. $12 + 87 + 18$; commutative property of addition

_____

4. 17×1; identity property of multiplication

_____

5. $\frac{1}{5} \times 5$; inverse property of multiplication

_____

6. $10 \times 12 \times 3$; commutative property of multiplication

_____

7. $12 \times (2 + 3)$; distributive property

_____

8. $(19 \times 5) \times 4$; associative property of multiplication

_____

Answer the questions about the way the following problem was solved.

1) $7(x + 4) = 3x$

2) $7x + 28 = 3x$

3) $-7x + 7x + 28 = -7x + 3x$

4) $28 = -4x$

5) $-\frac{1}{4} \times 28 = -\frac{1}{4} \times -4x$

6) $-7 = x$

9. On the right side of the equation, what property did the person use

to go from step (5) to step (6)? _____

10. Is $x = 7$ the correct solution? Prove your answer.

_____

Name _____ Date _____

Simplifying and Solving Equations: Lesson 1B

Practice

Apply the properties to simplify the expressions. Write the resulting expressions.

1. $8 \times (5 \times 13)$; associative property of multiplication

_____

2. $6 \times 19 \times 5$; commutative property of multiplication

_____

3. $-29 + 29$; inverse property of addition

_____

4. 143×1; identity property of multiplication

_____

5. $20 \times (5 + 1)$; distributive property

_____

6. $15 + 27 + 25$; commutative property of addition

_____

7. $8 \times \frac{1}{8}$; inverse property of multiplication

_____

8. $12 + (28 + 34)$; associative property of addition

_____

Answer the questions about the way the following problem was solved.

(1) $-3(x - 8) = x$

(2) $-3x + 24 = x$

(3) $3x - 3x + 24 = 3x + x$

(4) $24 = 4x$

(5) $\frac{1}{4} \times 24 = \frac{1}{4} \times 4x$

(6) $6 = x$

9. On the right side of the equation, what property did the person use

to go from step (5) to step (6)? _____

10. Is $x = 6$ the correct solution? Prove your answer.

_____ __ __ = __ __ __ =

Name _____ Date _____

Simplifying and Solving Equations: Lesson 2A

Practice

Write the solutions to the two-step equations. Verify your solutions by substituting for *x*. Show your work.

1. $5x - 2 = 18$ 　　　　　　　　　　　$5 \times \underline{\quad} - 2 = 18$

2. $x \div 4 - 3 = 3$ 　　　　　　　　　$\dfrac{\overline{\quad}}{4} - 3 = 3$

3. $-3x + 9 = 15$ 　　　　　　　　　$-3 \times \underline{\quad} + 9 = 15$

4. $12 = 2x - 4$ 　　　　　　　　　$12 = 2 \times \underline{\quad} - 4$

5. $3 = 7x + 3$ 　　　　　　　　　　$3 = 7 \times \underline{\quad} + 3$

6. $\frac{1}{2}x - 5 = 4$ 　　　　　　　　$\frac{1}{2} \times \underline{\qquad} - 5 = 4$

Simplifying and Solving Equations: Lesson 2B

Practice

Write the solutions to the two-step equations. Verify your solutions by substituting for *x*. Show your work.

1. $3x + 5 = 17$ $\qquad$ $3 \times \underline{\quad} + 5 = 17$

2. $-4x - 3 = 21$ $\qquad$ $-4 \times \underline{\quad} - 3 = 21$

3. $\frac{1}{3}x + 7 = 10$ $\qquad$ $\frac{1}{3} \times \underline{\quad} + 7 = 10$

4. $x \div 6 + 4 = 5$ $\qquad$ $\dfrac{\underline{\quad}}{6} + 4 = 5$

5. $10 = 6 + 4x$ $\qquad$ $10 = 6 + 4 \times \underline{\quad}$

6. $6 = 2x - 8$ $\qquad$ $6 = 2 \times \underline{\quad} - 8$

Name _____ Date _____

Simplifying and Solving Equations: Lesson 3A

Practice

Write the solutions to the two-step inequalities. Verify your solutions by substituting for *x*. Show your work.

1. $2x - 4 > 10$ 2_____ $- 4 > 10$

2. $3x + 5 < 17$ 3_____ $+ 5 < 17$

3. $21 \leq 6x + 9$ $21 \leq 6$_____ $+ 9$

4. $-10 > 2x + 6$ $-10 > 2$_____ $+ 6$

5. $11x + 1 \geq 12$ 11_____ $+ 1 \geq 12$

6. $51 < 4x + 7$ $51 < 4$_____ $+ 7$

Name _____ Date _____

Simplifying and Solving Equations: Lesson 3B

Practice

Write the solutions to the two-step inequalities. Verify your solutions by substituting for *x*. Show your work.

1. $3x - 8 < 13$ 3_____$- 8 < 13$

2. $64 < 5x + 9$ $64 < 5$_____$+ 9$

3. $7x + 3 > 31$ 7_____$+ 3 > 31$

4. $11 \leq 4x - 9$ $11 \leq 4$_____$- 9$

5. $28 > 6x - 8$ $28 > 6$_____$- 8$

6. $9x + 2 \geq 20$ 9_____$+ 2 \geq 20$

Name _____ Date _____

Simplifying and Solving Equations: Lesson 4A

Practice

Answer the following questions.

The Crumb and Coffee Shoppe sells 6 of its special muffins for $3.90. At this rate (price), how much would you pay for 10 muffins?

1. What is being asked? _____

2. What is the rate that you need? _____

3. How do you find this rate? _____

4. What is this rate? _____

5. How do you use this rate to find out how much 10 muffins cost?

_____

6. How much would 10 muffins cost? _____

A trucker drives 248 miles in 4 hours. At this rate how long would it take her to drive 682 miles?

7. What is her rate? _____

8. How did you find her rate? _____

9. How do you use this rate to find out how long it will take her to drive

682 miles? _____

10. At this rate, how long will it take her to drive 682 miles?

_____

Name _____ Date _____

Simplifying and Solving Equations: Lesson 4B

Practice

Answer the following questions.

Maria, Manuella's office assistant, can type 5 pages of financial reports in 2 hours. At this rate, how many pages of financial reports can she type in 5 hours?

1. What is being asked? _____

2. What is the rate that you need? _____

3. How do you find this rate? _____

4. What is this rate? _____

5. How do you use this rate to find out how many pages she can type in

5 hours? _____

6. How many pages can she type in 5 hours?

_____

Chef Andre can make 15 of his special gourmet salads in 12 minutes. At this rate, how long would it take him to prepare 40 of these salads?

7. What is his rate? _____

8. How did you find his rate? _____

9. How do you use this rate to find out how long it will take him to make

40 salads? _____

10. At this rate, how long will it take him to make 40 salads?

_____

Points in the Coordinate Plane: Lesson 1A

Practice

Plot each of the points on the grid provided. Identify the points by their coordinates.

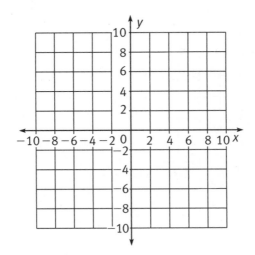

1. (2, 5)

2. (2, −4)

3. (0, −3)

4. (0, 6)

5. (−2, −6)

6. (7, 1)

7. (3, −5)

8. (5, 0)

9. (−3, 5)

10. (−5, −4)

11. Describe how to get from Point 1 above to Point 5. Be sure to describe left-right movement first and then up-down movement.

_____

12. Describe how to get from Point 10 above to Point 4.

_____

13. Describe how to get from Point 3 above to Point 4.

_____

14. Describe how to get from Point 7 above to Point 8. Be sure to describe left-right movement first and then up-down movement.

_____

Name _____ Date _____

Points in the Coordinate Plane: Lesson 1B

Practice

Plot each of the points on the grid provided. Identify the points by their coordinates.

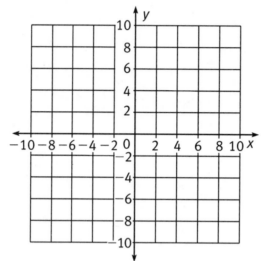

1. (0, 5)
2. (1, −5)
3. (4, −3)
4. (0, 6)
5. (−1, −6)

6. (4, 3)
7. (3, −4)
8. (2, 0)
9. (−3, 0)
10. (−6, −5)

11. Describe how to get from Point 4 above to Point 8. Be sure to describe left-right movement first and then up-down movement.

_____

12. Describe how to get from Point 1 above to Point 5.

_____

13. Describe how to get from Point 3 above to Point 6.

_____

14. Describe how to get from Point 8 above to Point 10.

_____

Name _____ Date _____

Points in the Coordinate Plane: Lesson 2A

Practice

Graph these points on a grid: A (−3, −1); B (−2, 2); C (3, 2); and D (7, −1). Connect the dots to make a geometric shape. Identify the shape you have made and answer the questions.

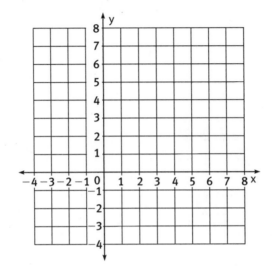

1. The shape is a _____

2. In which quadrant is point A located? _____

3. In which quadrant is point B located? _____

4. In which quadrant is point C located? _____

5. In which quadrant is point D located? _____

6. What is the relationship between side BC and side AD? _____

Write the quadrant in which each point is located. You can use a coordinate grid to help you identify the location. If the point is not in a quadrant, list the axis or origin on which it lies.

7. (2, 6) _____

8. (−5, 4) _____

9. (1, −10) _____

10. (−6, −9) _____

11. (5, 0) _____

12. (0, 0) _____

13. (−4, −3) _____

14. (0, −2) _____

Name _____ Date _____

Points in the Coordinate Plane: Lesson 2B

Practice

Graph these points on a grid: A (3, −1); B (1, 3); C (−4, 3); D (−5, 0) and E (0, −3). Connect the dots to make a geometric shape. Identify the shape you have made and answer the questions.

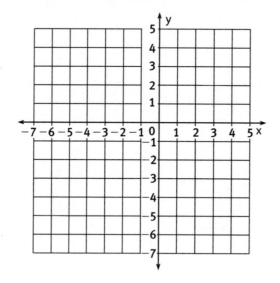

1. The shape is a _____

2. Describe the location of point A. _____

3. Describe the location of point B. _____

4. Describe the location of point C. _____

5. Describe the location of point D. _____

6. Describe the location of point E. _____

Write the quadrant in which each point is located. You can use a coordinate grid to help you identify the location. If the point is not in a quadrant, list the axis or origin on which it lies.

7. (−2, 5) _____

8. (0, 0) _____

9. (−1, −8) _____

10. (6, −3) _____

11. (−4, 0) _____

12. (−2, 2) _____

13. (2, 5) _____

14. (0, −7) _____

Name _____ Date _____

Points in the Coordinate Plane: Lesson 3A

Practice

Plot the given points on the coordinate grid. Use a ruler and draw a line that goes through the origin and the point you have placed on the grid.

1. (3, 1)

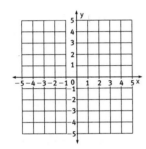

3. (−4, −1)

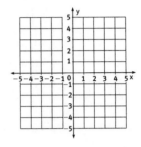

2. (4, 0)

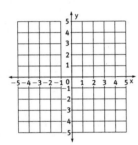

4. (2, −1)

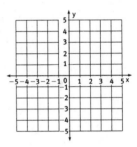

Plot the given points on the coordinate grid. Use a ruler and draw a line that goes through the two points you have placed on the grid.

5. (1, −3) and (−2, 1)

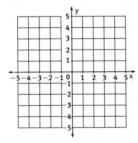

7. (−3, 4) and (0, 3)

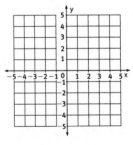

6. (0, 4) and (0, 1)

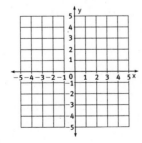

8. (−2, −4) and (2, 0)

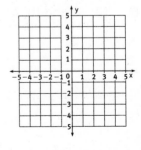

Name _____ Date _____

Points in the Coordinate Plane: Lesson 3B

Practice

Plot the given points on the coordinate grid. Use a ruler and draw a line that goes through the origin and the point you have placed on the grid.

1. $(-2, 3)$

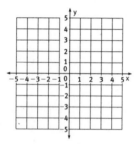

3. $(4, 1)$

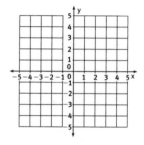

2. $(-3, 0)$

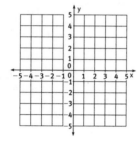

4. $(-2, -2)$

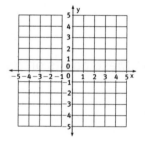

Plot the given points on the coordinate grid. Use a ruler and draw a line that goes through the two points you have placed on the grid.

5. $(1, 4)$ and $(2, -4)$

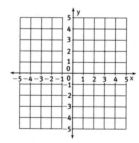

7. $(-2, 4)$ and $(0, 1)$

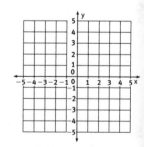

6. $(3, 4)$ and $(0, 4)$

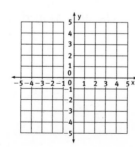

8. $(-1, -4)$ and $(3, 0)$

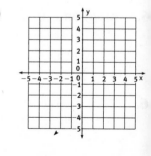

Name _____ Date _____

Points in the Coordinate Plane: Lesson 4A

Practice

1. Graph these points on the coordinate plane: A (5, 7); B (1, 7); C (1, 1); and D (5, 1). Then connect the points in alphabetical order. When you reach point D, stop. Identify the shape you have made.

The shape is a _____.

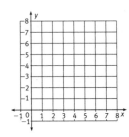

2. Graph these points on the coordinate plane: A (5, 7); B (1, 7); C (1, 1); D (5, 1); E (3, 4); and F (1, 4). Connect points A, B, C and D in alphabetical order. Stop and pick up your pencil. Now connect point E to point F. Stop. Identify the shape you have made.

The shape is a _____.

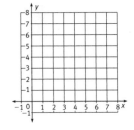

3. Graph these points on the coordinate plane: A (5, 1); B (5, 3); C (4, 4); D (5, 5); E (5, 7); F (1, 7); G (1, 4); H (1, 1). Connect points A, B, C, D, E, F, G, and H in alphabetical order. Connect point G to point C. Connect point A to point H. Stop. Identify the shape you have made.

The shape is a _____.

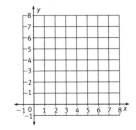

4. Graph these points on the coordinate plane: A (1, 1); B (3, 7); C (5, 1); D (2, 4); and E (4, 4). Connect points A, B, and C in alphabetical order. Stop and pick up your pencil. Now connect point D to point E. Stop. Identify the shape you have made.

The shape is a _____.

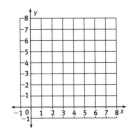

5. Graph these points on the coordinate plane: A (5, 1); B (5, 7); C (1, 1); D (1, 7); E (1, 4); and F (5, 4). Connect point A to point B. Connect point C to point D. Connect point F to point E. Stop. Identify the shape you have made.

The shape is a _____.

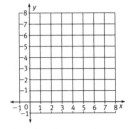

6. Rearrange your answers from Problems 1 through 5 to spell a very familiar word.

_____ _____ _____ _____ _____

Name _____ Date _____

Points in the Coordinate Plane: Lesson 4B

Practice

1. Graph these points on the coordinate plane: A (5, 7); B (1, 7); C (1, 1); and D (5, 1). Then connect the points in alphabetical order. When you reach point D, stop. Identify the shape you have made.

   The shape is a _____.

   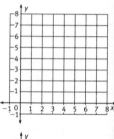

2. Graph these points on the coordinate plane: A (5, 7); B (1, 7); C (1, 1); and D (5, 1). Then connect the points in alphabetical order. When you reach point D, connect D to point A. Stop. Identify the shape you have made.

   The shape is a _____.

3. Graph these points on the coordinate plane: A (5, 1); B (5, 7); C (1, 1); and D (1, 7). Connect point A to point B. Connect point C to point D. Connect point D to point A. Stop. Identify the shape you have made.

   The shape is a _____.

   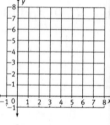

4. Graph these points on the coordinate plane: A (5, 7); B (1, 7); C (1, 1); D (5, 1); E(3,4); and F(1,4). Connect points A, B, C and D in alphabetical order. Stop and pick up your pencil. Now connect point E to point F. Stop. Identify the shape you have made.

   The shape is a _____.

   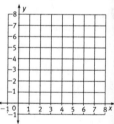

5. Graph these points on the coordinate plane: A (1, 1); B (3, 7); C (5, 1); D (2, 4); and E (4, 4). Connect points A, B, and C in alphabetical order. Stop and pick up your pencil. Now connect point D to point E. Stop. Identify the shape you have made.

   The shape is a _____.

   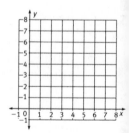

6. Rearrange your answers from Problems 1 through 5 to spell a very familiar word.

   _____ _____ _____ _____ _____

Name _____ Date _____

Line Properties: Lesson 1A

Practice

Determine the length of the line segments that connect these pairs of coordinates. Show your work.

1. (4, 2) and (8, 2)

_____

4. (1, −7) and (12, −7)

_____

2. (−6, 2) and (0, 2)

_____

5. (−5, 3) and (−2, 3)

_____

3. (0, 9) and (8, 9)

_____

6. (−4, −3) and (−9, −3)

_____

Determine the horizontal distance between the endpoints of the line segments that connect these pairs of coordinates. Show your work.

7. (5, 4) and (4, 2)

_____

10. (0, −7) and (6, −3)

_____

8. (−4, 6) and (−9, 4)

_____

11. (0, 7) and (−5, 4)

_____

9. (3, 9) and (−4, 6)

_____

12. (4, −1) and (12, −7)

_____

Line Properties: Lesson 1B

Practice

Determine the length of the line segments that connect these pairs of coordinates. Show your work.

1. (5, 3) and (9, 3)

_____

2. (−3, 5) and (−5, 5)

_____

3. (0, 6) and (3, 6)

_____

4. (8, 0) and (12, 0)

_____

5. (−7, 4) and (−2, 4)

_____

6. (0, 12) and (−6, 12)

_____

Determine the horizontal distance between the endpoints of the line segments that connect these pairs of coordinates. Show your work.

7. (7, −2) and (5, −1)

_____

8. (−5, −8) and (−12, −3)

_____

9. (−4, 6) and (9, −3)

_____

10. (0, 7) and (−6, 7)

_____

11. (2, 7) and (0, 4)

_____

12. (−2, −4) and (8, −6)

_____

Line Properties: Lesson 2A

Practice

Determine the length of the line segments that connect these pairs of coordinates. Show your work.

1. (4, 3) and (4, 7)

_____

2. (−3, 5) and (−3, 12)

_____

3. (0, 9) and (0, −7)

_____

4. (8, 0) and (8, 6)

_____

5. (−8, 6) and (−8, 3)

_____

6. (−6, −3) and (−6, 10)

_____

Determine the vertical distance between the endpoints of the line segments that connect these pairs of coordinates. Show your work.

7. (7, −2) and (5, −1)

_____

8. (−5, −6) and (−10, −3)

_____

9. (−4, 5) and (9, −3)

_____

10. (0, −4) and (−6, 7)

_____

11. (2, 7) and (3, 4)

_____

12. (−2, −4) and (8, −6)

_____

Name _____ Date _____

Line Properties: Lesson 2B

Practice

Determine the length of the line segments that connect these pairs of coordinates. Show your work.

1. (5, 5) and (5, 12)

_____

2. (−3, 5) and (−3, 8)

_____

3. (0, 6) and (0, −4)

_____

4. (8, 0) and (8, −5)

_____

5. (−7, 4) and (−7, 9)

_____

6. (0, 12) and (0, 4)

_____

Determine the vertical distance between the endpoints of the line segments that connect these pairs of coordinates. Show your work.

7. (7, −2) and (5, −7)

_____

8. (−2, −8) and (−12, −6)

_____

9. (−4, 6) and (9, −8)

_____

10. (0, 7) and (−6, −3)

_____

11. (2, 10) and (0, 4)

_____

12. (−3, 5) and (8, 9)

_____

Practice

Using the right triangle on the coordinate grid to answer the following questions. You will also use the Pythagorean theorem to find the length of the Hypotenuse.

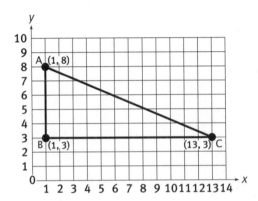

. Find the length of side AB. Call it *a* in the Pythagorean theorem. Show your work.

_____

2. Square the length of side AB. Show your work.

_____

3. Find the length of side BC. Call it *b* in the Pythagorean theorem. Show your work.

_____

4. Square the length of side BC. Show your work.

_____

5. Write the Pythagorean theorem.

_____

6. Substitute the values for a^2 and b^2.

_____

7. Simplify the equation you have written by combining the numbers.

_____

8. What number multiplied by itself gives 169?

_____

Therefore, the length of the hypotenuse, side

AC, in this right triangle is _____.

Use the Pythagorean theorem to find the length of the hypotenuse. Show your work below.

9. $a = 4, b = 3, c =$ _____

10 $a = 7, b = 24, c =$ _____

11. $a = 24, b = 10, c =$ _____

12. $a = 6, b = 8, c =$ _____

Name _____ Date _____

Line Properties: Lesson 3B

Practice

Use the right triangle on the coordinate grid to answer the following questions. You will also use the Pythagorean theorem to find the length of the hypotenuse.

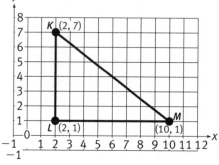

1. Find the length of side KL. Call it *a* in the Pythagorean theorem. Show your work.

   _____

2. Square the length of side KL. Show your work.

   _____

3. Find the length of side LM. Call it b in the Pythagorean theorem. Show your work.

   _____

4. Square the length of side LM. Show your work.

   _____

5. Write the Pythagorean theorem.

   _____

6. Substitute the values for a^2 and b^2.

   _____

7. Simplify the equation you have written by combining the numbers.

   _____

8. What number multiplied by itself gives 100?

   _____

Therefore, the length of the hypotenuse, side KM, in this right triangle is _____.

Given the following endpoints, solve for the length of the line segment. You may find it helpful to plot the two points and the line segment between them on a piece of graph paper and draw the right triangle for which this line segment is the hypotenuse.

9. (1, 5) and (5, 8)

   The length is _____.

10. (2, 0) and (9, 24)

    The length is _____.

11. (2, 2) and (7, 14)

    The length is _____.

12. (2, 4) and (8, 12)

    The length is _____.

Name _____ Date _____

Line Properties: Lesson 4A

Practice

Consider these groups of three numbers and determine whether they are the sides in a right triangle or not. Explain your answer. You will use the Pythagorean theorem. In each case the hypotenuse will be identified by c.

. 8 , 6, $c = 10$

_____

_____

4. 4, 4, $c = 32$

_____

_____

. 4, 9, $c = 12$

_____

_____

5. 6, 10, $c = 6$

_____

_____

. 3, 6, $c = 7$

_____

_____

6. 9, 8, $c = 12$

_____

_____

Find the missing leg or hypotenuse in each of these problems.

. $a = 3$, $c = 5$, find b.

$b =$ _____

10. $b = 15$, $c = 25$, find a.

$a =$ _____

8. $a = 12$, $b = 9$, find c.

$c =$ _____

11. $a = 14$, $c = 50$, find b.

$b =$ _____

. $b = 12$, $c = 13$, find a.

$a =$ _____

12. $a = 10$, $b = 24$, find c.

$c =$ _____

Line Properties: Lesson 4B

Practice

Consider these groups of three numbers and determine whether they are the sides in a right triangle or not. Explain your answer. You will use the Pythagorean theorem. In each case the hypotenuse will be identified by c.

1. 9, 12, $c = 15$

_____

_____

4. 5, 12, $c = 17$

_____

_____

2. 5, 10, $c = 12$

_____

_____

5. 0.6, 0.8, $c = 1$

_____

_____

3. 4, 5, $c = 6$

_____

_____

6. 3, 6, $c = 9$

_____

_____

Find the missing leg or hypotenuse in each of these problems.
Use numbers you know to approximate the missing side length.

7. $a = 5$, $b = 6$, find c.

b = between _____ and _____

10. $b = 12$, $c = 15$, find a.

a = between _____ and _____

8. $a = 7$, $b = 9$, find c.

c = between _____ and _____

11. $a = 12$, $c = 20$, find b.

b = _____

9. $b = 10$, $c = 12$, find a.

a = between _____ and _____

12. $a = 3$, $b = 5$, find c.

c = between _____ and _____

Name _____ Date _____

Change and Slope: Lesson 1A

Practice

These two coordinate grids show lines with "special" slopes. Calculate the slopes and then answer the questions.

1.

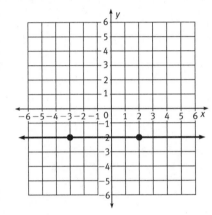

2.

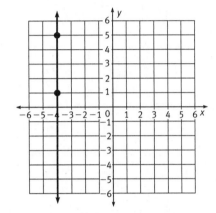

The slope of this line is $\frac{0}{5}$ or _____.

The slope of any horizontal line is _____.

The slope of this line is $\frac{4}{0}$ or _____.

The slope of any vertical line is _____.

Complete this table. Find the change in y-values and the change in the x-values for the pairs of points that are given. Then, calculate the slope.

	Points	Change in y-values	Change in x-values	Slope
3.	(2, 4) and (6, 5)	5 − 4 = 1	6 − 2 = 4	
4.	(−2, −2) and (−4, −4)			
5.	(3, 5) and (2, 7)		2 3 1	
6.	(4, −6) and (−5, 0)	0 − −6 = 6		
7.	(5, 2) and (−5, −2)			
8.	(2, −2) and (−4, 4)			
9.	(0, 0) and (1, 3)			
10.	(−2, 3) and (6, 0)			
11.	(4, 6) and (0, 0)			
12.	(3, −4) and (1, −4)			

Name _____ Date _____

Change and Slope: Lesson 1B

Practice

These two coordinate grids show lines with special slopes. Calculate the slopes and then answer the questions.

1.

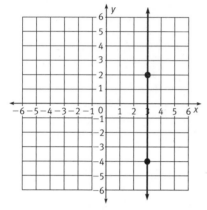

The slope of this line is $\frac{6}{0}$ or _____.

The slope of any vertical line is _____.

2.

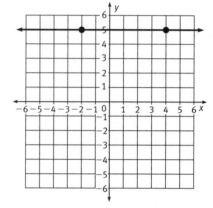

The slope of this line is $-\frac{0}{6}$ or _____.

The slope of any horizontal line is _____.

Complete this table. Find the change in *y*-values and the change in the *x*-values for the pairs of points that are given. Then, calculate the slope.

	Points	Change in *y*-values	Change in *x*-values	Slope
3.	$(-2, 4)$ and $(3, 5)$	$5 - 4 = 1$	$3 - -2 = 5$	
4.	$(-2, -3)$ and $(-4, -5)$			
5.	$(0, 5)$ and $(-2, 7)$		$-2 - 0 = -2$	
6.	$(-3, 5)$ and $(-4, 0)$	$0 - 5 = -5$		
7.	$(4, 2)$ and $(8, 4)$			
8.	$(2, 3)$ and $(-5, 4)$			
9.	$(0, 0)$ and $(2, 8)$			
10.	$(-1, 3)$ and $(5, 0)$			
11.	$(5, 12)$ and $(0, 0)$			
12.	$(7, -6)$ and $(1, -4)$			

Change and Slope: Lesson 2A

Practice

Find the slopes using the points given on the graphs. Remember that slope is the change in the *y*-values written over the change in the *x*-values ("rise over run").

1.

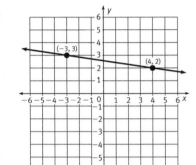

The slope of this line is _____.

2.

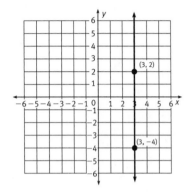

The slope of this line is _____.

3.

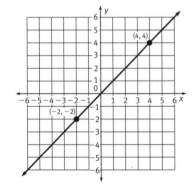

The slope of this line is _____.

4.

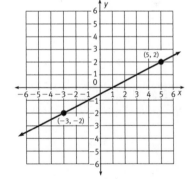

The slope of this line is _____.

Use the formula to calculate the slope of the line between the given points. Compare your answers to those above.

5. (3, 2) and (−3, 8)

The slope of this line is _____.

6. (1, 3) and (−2, −4)

The slope of this line is _____.

7. (9, 5) and (9, 2)

The slope of this line is _____.

8. (4, 1) and (7, −9)

The slope of this line is _____.

Change and Slope: Lesson 2B

Practice

Find each slope using the points given on the graphs. Remember that slope is the change in the *y*-values written over the change in the *x*-values ("rise over run").

1.

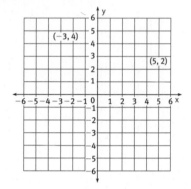

The slope of this line is _____.

3.

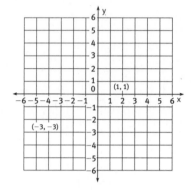

The slope of this line is _____

2.

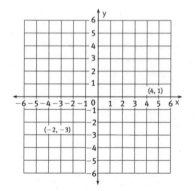

The slope of this line is _____.

4.

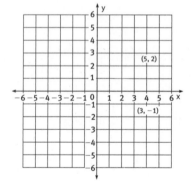

The slope of this line is _____

Use the formula to calculate the slope of the line between the given points. Compare your answers to those above.

5.   (1, 1) and (−5, 8)

   The slope of this line is _____.

7.   (9, 3) and (5, 3)

   The slope of this line is _____

6.   (−7, −6) and (2, 0)

   The slope of this line is _____.

8.   (6, 4) and (1, 0)

   The slope of this line is _____

Name _____ Date _____

Change and Slope: Lesson 3A

Practice

Graph the direct variation relationship 1 gallon = 4 quarts. Answer the questions to help you decide how the graph should look. Remember that two points determine a line. You will find a third point to make sure the other two points are correct.

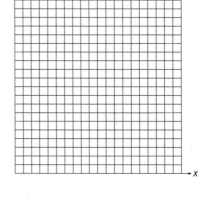

. Write two equations in x and y expressing this relationship.

_____

. Write three ordered pairs that satisfy this relationship.

_____, _____, and _____

. Which quantity will you graph on the x-axis? _____

. Which quantity will you graph on the y-axis? _____

. What scale could you use on the x-axis? _____

. What scale could you use on the y-axis? _____

. Graph the relationship.

. Looking at the line on the graph, what is the x-value when $y = 16$? _____

Here is the graph of the direct variation $y = -2x$. Looking at the graph, answer the following questions.

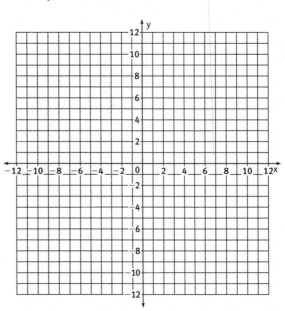

. When $x = 2$, what is the value of y? _____

0. When $y = 12$, what is the value of x? _____

1. When $y = 0$, what is the value of x? _____

2. When $x = -6$, what is the value of y? _____

3. When $x = 5$, what is the value of y? _____

4. When $y = -8$, what is the value of x? _____

Name _____ Date _____

Change and Slope: Lesson 3B

Practice

Graph the direct variation relationship 1 cup = 8 ounces. Answer y the questions to help you decide how the graph should look. Remember that two points determine a line. You will find a third point to make sure the other two points are correct.

1. Write two equations in x and y expressing this relationship.

   _____

2. Write three ordered pairs that satisfy this relationship.

   _____, _____, and _____

3. Which quantity will you graph on the x-axis? _____

4. Which quantity will you graph on the y-axis? _____

5. What scale could you use on the x-axis? _____

6. What scale could you use on the y-axis? _____

7. Graph the relationship.

8. Looking at the line on the graph, what is the x-value when $y = 16$? _____

Here is the graph of the direct variation $y = 5x$. Looking at the graph, answer the following questions.

9. When $x = 2$, what is the value of y? _____

10. When $y = -15$, what is the value of x? _____

11. When $y = 0$, what is the value of x? _____

12. When $x = -2$, what is the value of y? _____

13. When $x = 4$, what is the value of y? _____

14. When $y = -5$, what is the value of x? _____

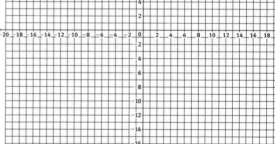

Change and Slope: Lesson 4A

Practice

The perimeter of a regular octagon, like a stop sign, can be found by adding the lengths of the 8 equal sides or by using the formula $P = 8s$, where s is the length of one of the sides. Answer the following questions and then graph this relationship.

. Write the formula as an equation involving x and y. _____

. Would you pick negative values for x or y? Why or why not?

_____

. Write three ordered pairs of numbers that can help you graph this line. Remember to keep the x-values fairly small so that the graph doesn't

get too large. _____

. What quantity will be indicated on the horizontal axes? _____

. What quantity will be indicated on the vertical axes? _____

. Plot the three points that you picked and draw the line through those points.

. From the points you graphed on the coordinate grid, calculate the slope of the line (change in y-values/change in x-values or rise/run).

_____

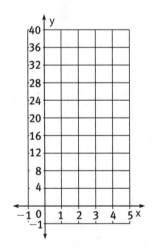

For each pair of points below, calculate the slope (ratio) of the line between them. You may want to plot the points on a graph to help you find the slope.

. (3, 5) and (4, 1)

_____

. (−2, −6) and (3, −2)

_____

10. (0, 4) and (5, −2)

_____

11. (3, 2) and (−5, 0)

_____

Name _____ Date _____

Change and Slope: Lesson 4B

Practice

The number of quarters in a given number of dollars can be described by the formula $Q = 4D$, where D is the number of dollars. Answer the following questions and then graph this relationship.

1. Write the formula as an equation involving x and y. _____

2. Would you pick negative values for x or y? Why or why not?

_____

3. Write three ordered pairs of numbers that can help you graph this line. Remember to keep the x values fairly small so that the graph doesn't

get too large. _____

4. What quantity will be indicated on the horizontal axes? _____

5. What quantity will be indicated on the vertical axes? _____

6. Plot the three points that you picked and draw the line through those points.

7. From the points you graphed on the coordinate grid, calculate the slope of the line (change in y-values/change in x-values or rise/run).

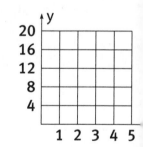

_____

For each pair of points below, calculate the slope (ratio) of the line between them. You may want to plot the points on a graph to help you find the slope.

8. (5, 2) and (2, 7)

_____

10. (6, 0) and (3, −1)

_____

9. (−1, 5) and (2, −4)

_____

11. (6, 1) and (0, −5)

_____

Rates and Products: Lesson 1A

Practice

Consider the equation $y = -2x + 4$. **Answer the following questions, and then graph the line that corresponds to this equation.**

1. What is the *y*-intercept of this line? _____

2. Write the coordinates of the *y*-intercept. _____

3. What is the slope of this line? _____

4. From the *y*-intercept the slope tells you to move down _____

   units and then to the _____ 1 unit to find another point on

   this line. The coordinates of this point are _____.

5. From the *y*-intercept you could also move up _____ units and

   then back to the left _____ unit to find a third point on this

   line. The coordinates of this point are _____.

6. Graph the line which corresponds to the equation $y = -2x + 4$.

7. Following the same process that you used to graph $y = -2x + 4$, graph the line that corresponds to the equation $y = \frac{2}{3}x - 1$.

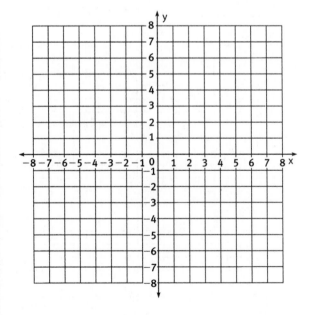

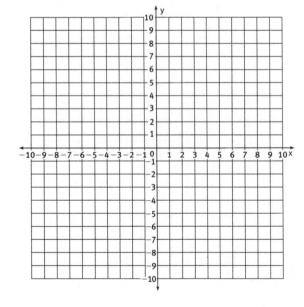

Name _____ Date _____

Rates and Products: Lesson 1B

Practice

Consider the equation $y = -3x + 1$. Answer the following questions, and then graph the line that corresponds to this equation.

1. What is the *y*-intercept of this line? _____

2. Write the coordinates of the *y*-intercept. _____

3. What is the slope of this line? _____

4. From the *y*-intercept the slope tells you to move down _____

   units and then to the _____ 1 unit to find another point on

   this line. The coordinates of this point are _____.

5. From the *y*-intercept you could also move up _____ units and

   then back to the left _____ unit to find a third point on this

   line. The coordinates of this point are _____.

6. Graph the line which corresponds to the equation $y = -3x + 1$.

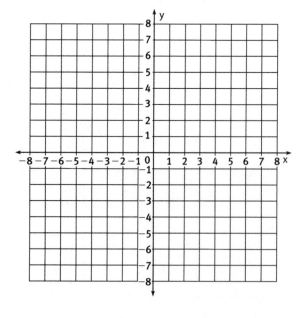

7. Following the same process that you used to graph $y = -3x + 1$, graph the line that corresponds to the equation $y = \frac{1}{4}x + 2$.

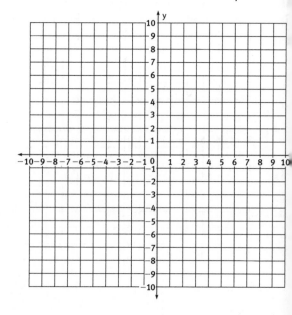

Name _____ Date _____

Rates and Products: Lesson 2A

Practice

The information contained in this chart shows the fuel efficiency for three cars used in a recent test. Find the rate (miles per gallon) for each car and then decide which car is the most fuel-efficient. (Hint: When finding miles per gallon in a test, you usually round to one decimal place in order to see a difference in the rate.)

Fuel Efficiency Test Results			
Vehicle	Miles	Gallons	Miles per Gallon (unit rate)
1. Car 1	250	8	
2. Car 2	240	7.6	
3. Car 3	248	8	

4. How would you write the rate per gallon for Car 3 rounded to the nearest tenth so you could compare it to the other two rates? _____

5. Which vehicle has the best fuel efficiency?

_____

6. Which vehicle has the worst fuel efficiency? _____

The information contained in this chart shows prices for three brands of trail mix in a local market. Find the rate (cost per ounce) for each brand and then decide which brand has the best price for trail mix. (Hint: When finding unit cost, you usually round to three decimal places in order to see a difference in the unit cost.)

Trail Mix Comparison			
Brand	Ounces	Price per Package	Cost per Ounce (unit rate)
7. Brand A	12	$1.86	
8. Brand B	15	$2.19	
9. Brand C	20	$2.95	

10. Describe the process you use to find the cost per ounce.

_____

_____

11. Which brand has the best price per ounce? _____

12. Which brand is the most expensive per ounce? _____

Name _____ Date _____

Rates and Products: Lesson 2B

Practice

A small packaging factory has three assembly lines that package sport cards for collectors. They package cards about football players, soccer players, baseball players, hockey players, and others. The information contained in this chart shows the production rate for the three lines. Find the rate (packages per hour) for each line and complete the information in the chart.

	Line	Packages	Hours	Packages per Hour (unit rate)
	Assembly Line Efficiency Results			
1.	Line 1	183	3	
2.	Line 2	145	2.5	
3.	Line 3	165	2.75	

4. Which Line is the most efficient? _____

5. Which Line is the least efficient? _____

6. At this rate, how many packages would Line 1 produce in an

8-hour day? _____

7. At this rate, how many packages would Line 2 produce in an

8-hour day? _____

8. At this rate, how many packages would Line 3 produce in an

8-hour day? _____

Rates and Products: Lesson 3A

Practice

Ki and Miguel have a small roofing company. They often get jobs working as subcontractors for companies who are building houses in subdivisions. Their team of 5 roofers can usually complete the roofs on 3 houses in 6 days.

1. How many person-days are needed for this 3-house project?

   _____

2. At this rate, how long would it take one roofer to complete the job if

   he or she had to work all alone? _____

3. At this rate, how much of the project does each roofer complete in

   one day? _____

4. How many days would it take if Ki and Miguel hired 6 roofers? _____

5. How many days would it take if they hired 10 roofers? _____

6. How many roofers would they need to hire to complete this job in

   2 days? _____

Tony and Betsy do the landscaping for this same developer. The two of them working together can finish the landscaping for a house in 10 hours.

7. How many person-hours are needed for a one-house project? _____

8. At this rate, how long would it take one of them to complete the job if

   he or she had to work all alone? _____

9. At this rate, how much of the project does each of them complete in

   one hour? _____

10. How many hours would it take if Betsy and Tony hired another worker? _____

11. How many hours would it take if they hired 2 more workers? _____

12. How many workers would they need to hire to complete this job in

    4 hours? _____

Rates and Products: Lesson 3B

Practice

Perry and Tammy have a small painting company. They often get jobs working as subcontractors for companies who are building houses in subdivisions. Their team of 4 painters can usually complete the inside painting on 2 houses in 5 days.

1. How many person-days are needed for this 2-house project? _____

2. At this rate, how long would it take one painter to complete the job if he or she had to work all alone? _____

3. At this rate, how much of the project does each painter complete in one day? _____

4. How many days would it take if Perry and Tammy hired 5 painters? _____

5. How many days would it take if they only had 2 painters? _____

6. How many painters would they need to hire to complete this 2-house project in 2 days? _____

Luis is a subcontractor who clears the land for this same developer. Luis's crew of 3 workers can clear the land for 2 houses in 8 hours.

7. How many person-hours are needed for a 2-house clearing? _____

8. At this rate, how long would it take one of them to complete the job if he or she had to work all alone? _____

9. At this rate, how much of the project does each of them complete in one hour? _____

10. How many hours would it take if Luis hired another worker? _____

11. How many hours would it take if he hired 3 more workers? _____

12. How many workers would he need to use to complete this job in 6 hours? _____

Name _____ Date _____

Rates and Products: Lesson 4A

Practice

The average speed at a recent Indianapolis 500 Mile Race was 206 miles per hour. At this rate, how many feet per minute does a racecar travel?

1. Write 206 miles per hour as a ratio (in fraction form) but keep the units with the numbers.

_____

2. Make a multiplication expression using all 3 ratios. Be sure to include the labels with the numbers.

_____

3. Use Dimensional Analysis to cancel out the units. Since the original question was feet per minute, you want feet remaining in the top and minutes in the bottom.

_____

4. Now finish the mathematics by multiplying across the numerators and denominators, reducing first as you go, or reduce the final result.

_____

5. Finally, answer the question using the correct labels.

_____

Find the error in this solution. Then, give the correct answer.

6. Change 150 feet per second to miles per hour.

$$\frac{150 \text{ ft}}{1 \text{ sec}} \times \frac{60 \text{ sec}}{1 \text{ hour}} \times \frac{1 \text{ mi}}{5,280 \text{ ft}} = \frac{150 \times 60 \times 1}{1 \times 1 \times 5,280}$$

$$= \frac{9000}{5280} = 1.705 \text{ miles per hour}$$

_____

The correct answer is _____.

Name _____ Date _____

Rates and Products: Lesson 4B

Practice

Hermani, a major league baseball pitcher, throws his fastball at an average rate of 94 miles per hour. How fast would that pitch be in feet per second?

1. Write 94 miles per hour as a ratio (in fraction form) but keep the units with the numbers.

_____

2. Make a multiplication expression using all 4 ratios. Be sure to include the labels with the numbers.

_____

3. Use Dimensional Analysis to cancel out the units. Since the original question was feet per second, you want feet remaining in the top and seconds in the bottom.

_____

4. Now finish the mathematics by multiplying across the numerators and denominators, reducing first as you go, or reduce the final result.

_____

5. Finally, answer the question using the correct labels.

_____

Find the error in this solution. Then, give the correct answer.

6. Convert $4.98 per quart to cost per ounce.

$$\frac{\$4.98}{1 \text{ quart}} \times \frac{1 \text{ quart}}{2 \text{ pints}} \times \frac{1 \text{ pint}}{2 \text{ cups}} \times \frac{1 \text{ cup}}{16 \text{ ounces}} = \frac{\$4.98 \times 1 \times 1 \times 1}{1 \times 2 \times 2 \times 16}$$

$$= \frac{4.98}{64} = \$0.078 \text{ per ounce}$$

_____

The correct answer is _____.

Name _____ Date _____

Manipulating Equations: Lesson 1A

Practice

Write *opposite* or *reciprocal* on the blank to make each statement true.

1. 5 is the _____ of $\frac{1}{5}$.

5. $\frac{1}{4}$ is the _____ of 4.

2. -8 is the _____ of 8.

6. $-\frac{1}{4}$ is the _____ of $\frac{1}{4}$.

3. $\frac{2}{3}$ is the _____ of $-\frac{2}{3}$.

7. $-\frac{5}{7}$ is the _____ of $-\frac{7}{5}$.

4. $\frac{2}{3}$ is the _____ of $\frac{3}{2}$.

8. 1 is the _____ of 1.

	Number	Opposite	Reciprocal
9.	9	-9	$\frac{1}{9}$
10.	-6		
11.	$\frac{3}{5}$		
12.	$\frac{1}{2}$		
13.	$-\frac{5}{8}$		
14.	$\frac{1}{3}$		
15.	-1		

Manipulating Equations: Lesson 1B

Practice

Draw a solid line from the number to its opposite or a dashed line from the number to its reciprocal.

1. $\frac{5}{8}$

2. -6

3. $\frac{1}{4}$

4. $\frac{-2}{3}$

5. 12

6. $\frac{1}{5}$

A. $\frac{1}{12}$

B. $\frac{-5}{8}$

C. $\frac{-1}{5}$

D. $\frac{2}{3}$

E. $\frac{-1}{6}$

F. 4

	Number	Opposite	Reciprocal
7.	$\frac{1}{3}$		
8.	10		
9.	$\frac{3}{4}$		
10.	-7		
11.	$-\frac{3}{5}$		
12.	6		

Manipulating Equations: Lesson 2A

Practice

Draw a line from the expression in Column 1 to the simplified expression in Column 2 when the distributive property has been corrrectly applied.

1. $2(x + 6)$

2. $-3(x + 4)$

3. $6(2x - 1)$

4. $-4(3x - 2)$

5. $5(x - 4)$

6. $3(2x - 3)$

A. $5x - 20$

B. $12x - 6$

C. $-12x + 8$

D. $6x - 9$

E. $2x + 12$

F. $-3x - 12$

7. Is $-2(x + 4) = -2x - 8$ a true or false statement? Prove whether this application of the distributive property is true or false by letting $x = 10$. Show your work. If the application is false, give the correct simplification.

8. Is $9x - 6 \quad 3 = 3x - 3$ a true or false statement? Prove whether this application of the distributive property is true or false by letting $x = 5$. Show your work. If the application is false, give the correct simplification.

Manipulating Equations: Lesson 2B

Practice

1. Is $4(3x + 1) = 12x + 1$ a true or false statement? Prove whether this application of the distributive property is true or false by letting $x = 2$. Show your work. If the application is false, give the correct simplification.

2. Is $-2(x - 4) = -2x + 8$ a true or false statement? Prove whether this application of the distributive property is true or false by letting $x = 6$. Show your work. If the application is false, give the correct simplification.

Simplify each of the following expressions by correctly applying the distributive property.

3. $6(x - 7)$

_____

4. $(-9x - 18) \div 9$

_____

5. $(8x - 24) \div 4$

_____

6. $7(-3x + 2)$

_____

7. $-4(2x + 1)$

_____

8. $-5(7 + 3x)$

_____

9. $(15x - 20) \div 5$

_____

10. $(-12x + 6) \div -2$

_____

Name _____ Date _____

Manipulating Equations: Lesson 3A

Practice

Write the inverse of each given equation.

1. If $\sqrt{49} = 7$, then _____.

2. If $6^4 = 1,296$, then _____.

3. If $3^5 = 243$, then _____.

4. If $\sqrt[3]{64} = 4$, then _____.

5. If $9^2 = 81$, then _____.

6. If $\sqrt[4]{10,000} = 10$, then _____.

Use your calculator to find the roots of these numbers. Round your answers to the nearest thousandth.

7. $\sqrt[3]{12} =$ _____

8. $\sqrt[4]{1,000} =$ _____

9. $\sqrt[5]{85} =$ _____

10. $\sqrt{29} =$ _____

Simplify each variable expression.

11. $\sqrt[3]{(64x^3)} =$ _____

12. $\sqrt[4]{(81x^8)} =$ _____

Manipulating Equations: Lesson 3B

Practice

Write the inverse of each given equation.

1. If $\sqrt{81} = 9$, then _____.

2. If $5^3 = 125$, then _____.

3. If $2^8 = 256$, then _____.

4. If $\sqrt[4]{1,296} = 6$, then _____.

5. If $12^3 = 1,728$, then _____.

6. If $\sqrt[5]{100,000} = 10$, then _____.

Use your calculator to find the roots of these numbers. Round your answers to the nearest thousandth.

7. $\sqrt[4]{100} =$ _____

8. $\sqrt[2]{13} =$ _____

9. $\sqrt[3]{30} =$ _____

10. $\sqrt[5]{250} =$ _____

Simplify each variable expression.

11. $\sqrt[6]{(x^{12})} =$ _____

12. $\sqrt[3]{(125x^9)} =$ _____

Practice

Draw a line from the power expression in Column 1 to its simplification in Column 2.

Column 1

1. 7^2

2. $64^{\frac{1}{2}}$

3. 6^3

4. $125^{\frac{1}{3}}$

5. $(-3x)^3$

6. $(2x)^5$

7. $(36x^2)^{\frac{1}{2}}$

8. $(49x^4y^4)^{\frac{1}{2}}$

9. $(-2x)^4$

10. $(11x)^2$

Column 2

A. $6x$

B. $16x^4$

C. $32x^5$

D. 49

E. $7x^2y^2$

F. $121x^2$

G. $-27x^3$

H. 216

I. 5

J. 8

Practice

Draw a line from the power expression in Column 1 to its simplification in Column 2.

Column 1

1. 4^3

2. $4^{\frac{1}{2}}$

3. 8^3

4. $64^{\frac{1}{3}}$

5. $(-5x)^3$

6. $(3x)^2$

7. $(64x^4)^{\frac{1}{2}}$

8. $(8x^3y^6)^{\frac{1}{3}}$

9. $(-4x)^4$

10. $(10x^2)^2$

Column 2

A. $2xy^2$

B. $100x^4$

C. $-125x^3$

D. 64

E. 512

F. $256x^4$

G. 2

H. 4

I. $9x^2$

J. $8x^2$

Practice

Use the Pythagorean theorem to determine whether each group of three numbers form a right triangle. Explain your answer. The hypotenuse equals c.

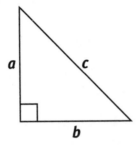

1. 8, 6, $c = 10$

_____

2. 2, 4, $c = 5$

_____

3. $\sqrt{10}$, 4, $c = \sqrt{26}$

_____

4. 10, 24, $c = 26$

_____

Find the missing leg or hypotenuse. If your answers are not perfect squares, leave the answers in square root form.

5. $a = 12$, $b = 5$, find c.

_____

8. $b = \sqrt{8}$, $c = \sqrt{24}$, find a.

_____

6. $a = 36$, $c = 60$, find b.

_____

9. $a = 5$, $c = 9$, find b.

_____

7. $b = 9$, $c = 15$, find a.

_____

10. $a = 24$, $c = 25$, find b.

_____

Name _____ Date _____

Exponents and Applications: Lesson 1B

Practice

Use the Pythagorean theorem to determine whether each group of three numbers form a right triangle. Explain your answer. The hypotenuse equals *c*.

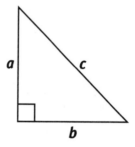

1. 16, 12, *c* = 20

_____

2. 4, 8, *c* = 10

_____

3. $\sqrt{8}$, 3, *c* = $\sqrt{17}$

_____

4. 7, 4, *c* = 8

_____

Find the missing leg or hypotenuse. If your answers are not perfect squares, leave the answers in square root form.

5. *a* = 8, *b* = 6, find *c*.

_____

6. *a* = 6, *c* = 9, find *b*.

_____

7. *b* = 10, *c* = 26, find *a*.

_____

8. *b* = $\sqrt{11}$, *c* = 6, find *a*.

_____

9. *a* = 21, *c* = 75, find *b*.

_____

10. *a* = 24, *b* = 18, find *c*.

_____

Practice

State whether the following statements are *true* or *false*. If the statement is false, write the correct answer.

1. $3^2 = 9$

_____

5. $4^0 = 4$

_____

2. $3^{-2} = \frac{1}{9}$

_____

6. $3^1 = 3$

_____

3. $4^3 = 12$

_____

7. $3^4 = 81$

_____

4. $4^{-1} = -4$

_____

8. $4^{-3} = -12$

_____

Simplify. Assume all variables do not equal 0.

9. $(2x)^1 =$ _____

12. $(4x^3)^{-1} =$ _____

10. $(4x)^3 =$ _____

13. $(6x)^2 =$ _____

11. $(-4x)^0 =$ _____

14. $(-2xy^2)^0 =$ _____

Name _____ Date _____

Exponents and Applications: Lesson 2B

Practice

State whether the following statements are *true* or *false*. If the statement is false, write the correct answer.

1. $7^3 = 343$

_____

2. $1^{-7} = -7$

_____

3. $7^{-3} = -343$

_____

4. $7^{-1} = \frac{1}{7}$

_____

5. $7^0 = 7$

_____

6. $7^4 = 2{,}401$

_____

7. $7^{-2} = \frac{1}{49}$

_____

8. $7^2 = 14$

_____

Simplify. Assume all variables do not equal 0.

9. $(-3x)^2 =$ _____

10. $(-x)^{-2} =$ _____

11. $(5xy)^0 =$ _____

12. $(-5x)^1 =$ _____

13. $(3xy)^{-1} =$ _____

14. $(6xy)^0 =$ _____

Exponents and Applications: Lesson 3A

Practice

tate whether the following statements are *true* or *false*. If the statement
false, write the correct answer. Leave the answers in exponential form.

. $4^3 \times 4^4 = 4^7$

_____

7. $(3^4)^3 = 3^{12}$

_____

. $(4x)^3 = 64x^3$

_____

8. $(3x^2)^3 = 9x^6$

_____

. $3^8 \div 3^4 = 3^2$

_____

9. $\left(\frac{3}{4}\right)^3 = \frac{27}{64}$

_____

. $(3x)^4 \times (3x)^2 = (3x)^8$

_____

10. $\left(\frac{x}{4}\right)^2 = \frac{x^2}{16}$

_____

. $4^2 \times 4^6 = 4^{12}$

_____

11. $4^6 \div 4^3 = 4^2$

_____

. $(4xy)^6 \div (4xy)^3 = (4xy)^3$

_____

12. $(4xy)^3 \times (4xy)^4 = (4xy)^{12}$

_____

Simplify. Assume all variables do not equal 0.

13. $(-3x)^0 =$ _____

16. $(3xy)^5 \div (3xy)^3 =$ _____

14. $(4x)^3 \times (4x)^3 =$ _____

17. $(-3xy)^1 =$ _____

15. $(4xy)^2 =$ _____

18. $\left(\frac{3}{xy}\right)^3 =$ _____

Exponents and Applications: Lesson 3B

Practice

State whether the following statements are *true* or *false*. If the statement is false, write the correct answer. Leave the answers in exponential form. Assume all variables are not equal to zero.

1. $5^3 \times 5^2 = 5^5$

_____

2. $(5x)^2 = 25x^2$

_____

3. $5^{10} \div 5^2 = 5^5$

_____

4. $(-3x)^4 = -81x^4$

_____

5. $(2^3)^2 = 2^5$

_____

6. $(8xy)^6 \div (8xy)^5 = (8xy)^1$, or $8xy$

_____

7. $\left(\frac{5}{2}\right)^4 = 5^{\frac{4}{2}}$

_____

8. $(5x^2) \times (5x^2)^3 = (5x^2)^4 = 5x^8$

_____

9. $5^4 \div 5^1 = 5^3$

_____

10. $\left(\frac{x^2}{5}\right)^2 = \frac{x^4}{5^2}$, or $\frac{x^4}{25}$

_____

11. $2^4 \times 2^4 = 4^4$

_____

12. $(-2xy)^5 \times (-2xy)^4 = (-2xy)^{20}$

_____

Simplify. Assume all variables do not equal 0.

13. $(6x^2)^1 = $ _____

14. $\left(\frac{2xy}{3}\right)^2 = $ _____

15. $(-2xy)^0 = $ _____

16. $(6xy)^4 \div (6xy)^3 = $ _____

17. $(6xy) \times (6xy)^2 = $ _____

18. $(2x^2)^4 = $ _____

Exponents and Applications: Lesson 4A

Name _____ Date _____

Practice

Each of the following simplifications has an error. Find the error. Explain what should have been done. Then simplify the expression following the Basic Rules. Assume all variables are not equal to zero.

$$x^5 \times x^5 = x^{25}$$

_____

The correct simplification is _____.

$$(-4x^2)(2x^2) = -8x^2$$

_____

The correct simplification is _____.

$$\frac{x^{10}}{x^2} = x^5$$

_____

The correct simplification is _____.

$$(-3x^2)^4 = -3x^8$$

_____

The correct simplification is _____.

Simplify each expression using the 5 Basic Rules, PEMDAS, the distributive property, and the rules for signed numbers. Assume all variables do not equal 0.

$x^3 \times x^6 =$ _____

9. $3x^4 \times 5x^2 =$ _____

$-4x^2(-3x + 1) =$ _____

10. $(-5)^{-2} \times (-5)^5 =$ _____

$\left(\frac{3}{5}\right)^3 =$ _____

11. $(4x)^0 \times (-2x)^2 =$ _____

$24x^3 \div -2x^5 =$ _____

12. $9x^4 - 4x^3 =$ _____

Name _____ Date _____

Exponents and Applications: Lesson 4B

Practice

Each of the following simplifications has an error. Find the error. Explain what should have been done. Then simplify the expression correctly following the 5 Basic Rules. Assume all variables are not equal to zero.

1. $-4x^5 \times -4x^5 = -8x^5$

_____

The correct simplification is _____

2. $14x^8 \div 7x^4 = 2x^2$

_____

The correct simplification is _____

3. $x^{-6} \div x^{-2} = x^{12}$

_____

The correct simplification is _____

4. $(5x^2)^3 = 15x^6$

_____

The correct simplification is _____

Simplify each expression using the 5 Basic Rules, PEMDAS, the distributive property, and the rules for signed numbers. Assume all variables do not equal 0.

5. $x^4 \times x^{-3} =$ _____

6. $-3x^2(8x - 1) =$ _____

7. $\left(\dfrac{2x}{7}\right)^2 =$ _____

8. $-6x^5 \div -3x^2 =$ _____

9. $-2x^2 \times 8x^2 =$ _____

10. $(-4)^{-3} \times (-4)^2 =$ _____

11. $(3x)^0 \times 2x =$ _____

12. $-5x^3 - 4x^5 =$ _____

Name _____ Date _____

Variables in Equations: Lesson 1A

Practice

Draw a line from the expression in Column 1 to its simplification Column 2.

$5x^2 + 7x^2$ **A.** $30x^3$

$(-3x^2)(-5x^2)$ **B.** $-5x^4$

$6x^4 - 9x^4$ **C.** $-8x^6$

$7x^2 + 4x - 5 + 6x^2 - 8$ **D.** $15x^4$

$120x^5 \div 4x^2$ **E.** $12x^2$

$(3x^2 - 5x^2)^3$ **F.** $13x^2 + 4x - 13$

Simplify. **Assume all variables do not equal 0.**

$8x^2 - 11x^2 + 5x =$ _____

10. $\frac{1}{4}(8x^3 - 12x^2 - 20x) =$ _____

$-2x(x - 6) + 4x^2 =$ _____

11. $(-2x^2)(5x^2) + 8x^3 =$ _____

$-100x^6 \div 4x^4 =$ _____

12. $3x(2x + 1) - 8x$ _____

Variables in Equations: Lesson 1B

Practice

Draw a line from the expression in Column 1 to its simplification in Column 2.

1. $9x^4 - 12x^4$ **A.** $-4x^6$

2. $40x^8 \div -10x^2$ **B.** $50x^6$

3. $(-5x^2)(-10x^4)$ **C.** $11x^3 - 14x$

4. $\frac{1}{2}(6x^3 - 4x^2 + 12x - 8)$ **D.** $3x^2$

5. $8x^3 - 5x + 3x^3 - 9x$ **E.** $-3x^4$

6. $(-3x)^2 + 6x^2$ **F.** $3x^3 - 2x^2 + 6x - 4$

Simplify. Assume all variables do not equal 0.

7. $3x - (-6x) - 4x^2 - 7x =$ _____

10. $12x^2 - 18x^3 \div 9x + 3 =$ _____

8. $(4x^3 - 6x^3)^2 =$ _____

11. $8x + (-3x)(2x) - 10x =$ _____

9. $150x^6 \div -25x^2 =$ _____

12. $15x(3x^2 - 5)$ _____

Variables in Equations: Lesson 2A

Practice

The following equations have been solved for you. After each step, write the property, process, or rule that was used to get to that form of the equation. Two of the more common processes are *combine like terms* and *simplify arithmetic*.

$$-4a - 12 = 3a + 9$$

$$-4a - 12 - 3a = 3a + 9 - 3a \qquad \underline{\hspace{4cm}}$$

$$-7a - 12 = 9 \qquad \underline{\hspace{4cm}}$$

$$-7a - 12 + 12 = 9 + 12 \qquad \underline{\hspace{4cm}}$$

$$-7a = 21 \qquad \underline{\hspace{4cm}}$$

$$\frac{-7a}{-7} = \frac{21}{-7} \qquad \underline{\hspace{4cm}}$$

$$a = -3 \qquad \underline{\hspace{4cm}}$$

Solve. Write each step, and be prepared to explain the property, rule, or process that you used.

$w - 4 = 10$ _____

$-6 = y + 5$ _____

$-21 = -3x$ _____

0. $\frac{2}{5}a = 12$ _____

11. $-3a - 1.4 = 5.2$ _____

12. $\frac{d}{5} = -10$ _____

13. $-4m + 6 = -14$ _____

14. $6c - 8c + 7 = -9$ _____

Variables in Equations: Lesson 2B

Practice

The following equations have been solved for you. After each step, write the property, process, or rule that was used to get to that form of the equation. Two of the more common processes are *combine like terms* and *simplify arithmetic.*

$$7m - 4m - 20 = -6 + 10$$

1.　　　$3m - 20 = -6 + 10$　_____

2.　　　$3m - 20 = 4$　_____

3.　$3m - 20 + 20 = 4 + 20$　_____

4.　　　　$3m = 24$　_____

5.　　　$\dfrac{3m}{3} = \dfrac{24}{3}$　_____

6.　　　　$m = 8$　_____

Solve. Write each step, and be prepared to explain the property, rule, or process that you used.

7. $h - 9 = 10$ _____

8. $\dfrac{2}{3}y = -12$ _____

9. $\dfrac{m}{-3} = 12$ _____

10. $132 = 8y + 3y$ _____

11. $-5k + 2.6 = 1.1$ _____

12. $t + 3t - 5 = 11$ _____

13. $-8d + 6 = 3d - 5$ _____

14. $7n - 2n + 10 = 4n$ _____

Name _____ Date _____

Variables in Equations: Lesson 3A

Practice

There is an error in each simplification. Explain what the error is and **s**implify the expression correctly.

. $-5(2a - 4) = -10a - 20$

The error is _____.

The correct simplification is _____.

. $(3x)^2 \times 2x = 3x^2 \times 2x = 6x^3$

The error is _____.

The correct simplification is _____.

There is an error in the solution for each equation. Explain what the **e**rror is and solve the equation correctly.

. $-8m + 6m + 3m = 5 + 4$
$-8m + 9m = 5 + 4$
$-1m = 9$
$\dfrac{-1m}{-1} = \dfrac{9}{-1}$
$m = -9$

The error is _____

_____.

The correct solution is _____.

4. $-4(2a + 6) - 10 = 6 - 8$
$-8a + 24 - 10 = 6 - 8$
$-8a + 14 = -2$
$-8a + 14 - 14 = -2 - 14$
$-8a = -16$
$\dfrac{-8a}{-8} = \dfrac{-16}{-8}$
$a = 2$

The error is _____

_____.

The correct solution is _____.

Solve each equation. Show your steps.

. $5y - 7y - 3y = 3^2 + 11$

$y =$ _____

. $-3(2a - 1) + 6 = 3a$

$a =$ _____

7. $4(p - 3) + 2(3p - 1) = 5 + 1$

$p =$ _____

8. $\frac{2}{5}(10d - 15) + 3d = 3^3 - 5$

$d =$ _____

Name _____ Date _____

Variables in Equations: Lesson 3B

Practice

There is an error in each simplification. Explain what the error is and simplify the expression correctly.

1. $-4d(-3d + 4d^2) = -4d(d^2) = -4d^3$

The error is _____

The correct simplification is _____

2. $-5a(6a^2 - 3a + 4) = -30a^3 + 15a - 20a = -30a^3 - 5a$

The error is _____

The correct simplification is _____

There is an error in the solution for each equation. Explain what the error is and solve the equation correctly.

3.
$$3m - 7m + 2m - 4 = 12$$
$$-4m + 2m - 4 = 12$$
$$2m - 4 = 12$$
$$2m = 16$$
$$m = 8$$

The error is _____

_____.

The correct solution is _____.

4.
$$5(3a - 4) = 4^2$$
$$15a - 4 = 4^2$$
$$15a - 4 = 16$$
$$15a = 20$$
$$\frac{15a}{15} = \frac{20}{15}$$
$$a = \frac{4}{3}$$

The error is _____

_____

The correct solution is _____

Solve each equation. Show your steps.

5. $6t - 4t + 3t - t = 3^2 - 1$

$t =$ _____

7. $4(2n - 5) - 3(n + 2) = 8 - 9$

$n =$ _____

6. $-5(2c - 3) + 4c = 18$

$c =$ _____

8. $2w - 7 = 6w - 5$

$w =$ _____

Name _____ Date _____

Variables in Equations: Lesson 4A

Practice

This equation has been solved and the solution has been proved. Write the explanation for what was done to reach each step.

Solution: **Explanation:**

$$-5(3t - 4) + 10 = 5^2$$

1. $-15t + 20 + 10 = 5^2$ _____

2. $-15t + 30 = 5^2$ _____

3. $-15t + 30 = 25$ _____

4. $-15t + 30 - 30 = 25 - 30$ _____

5. $-15t = -5$ _____

6. $\dfrac{-15t}{-15} = \dfrac{-5}{-15}$ _____

7. $t = \dfrac{1}{3}$ _____

Proof:

$$-5(3t - 4) + 10 = 5^2$$

8. $-5[3\left(\dfrac{1}{3}\right) - 4] + 10 = 5^2$ _____

9. $-5[1 - 4] + 10 = 5^2$ _____

10. $-5(-3) + 10 = 5^2$ _____

11. $15 + 10 = 5^2$ _____

12. $25 = 5^2$ _____

13. $25 = 25$ _____

Solve each equation, and prove your answer. Show your work.

14. $4a - 6a + 8a = 5 \times 8 + 2$

$a =$ _____

15. $-3(2t - 5) + 2(t - 8) = 3$

$t =$ _____

Variables in Equations: Lesson 4B

Practice

This equation has been solved and the solution has been proved. Write the explanation for what was done to reach each step.

Solution: **Explanation:**

$3(-2a + 4) + 4a - 8 = 3^2 + 1$

1. $-6a + 12 + 4a - 8 = 3^2 + 1$ _____

2. $-2a + 4 = 3^2 + 1$ _____

3. $-2a + 4 = 9 + 1$ _____

4. $-2a + 4 = 10$ _____

5. $-2a + 4 - 4 = 10 - 4$ _____

6. $-2a = 6$ _____

7. $\frac{-2a}{-2} = -\frac{6}{2}$ _____

8. $a = -3$ _____

Proof:

$3(-2a + 4) + 4a - 8 = 3^2 + 1$

9. $3[-2(-3) + 4] + 4(-3) - 8 = 3^2 + 1$ _____

10. $3[6 + 4] - 12 - 8 = 3^2 + 1$ _____

11. $3(10) - 12 - 8 = 3^2 + 1$ _____

12. $30 - 12 - 8 = 3^2 + 1$ _____

13. $30 - 12 - 8 = 9 + 1$ _____

14. $10 = 10$ _____

Solve each equation, and prove your answer. Show your work.

15. $5y + 3y - 2y + 8 = -16$

   $y = $ _____

16. $3(2m - 4) - 2(m + 1) = 3^2 + 1$

   $m = $ _____

Name _____ Date _____

Equations and Inequalities: Lesson 1A

Practice

Translate each of these sentences into an equation. Solve the equation. To prove that your answer is correct, read the problem with the answer in place. If it makes a true statement, write *true*.

1. The difference between a number and 7 is 6. Find the number.

   Equation: _____

   Answer: _____

   Proof: _____

   _____

4. The quotient of a number and 6 is 12. Find the number.

   Equation: _____

   Answer: _____

   Proof: _____

   _____

2. The product of −4 and a number is 28. Find the number.

   Equation: _____

   Answer: _____

   Proof: _____

   _____

5. Five less than three times a number is 22. Find the number.

   Equation: _____

   Answer: _____

   Proof: _____

   _____

3. The sum of −3 and a number is −8. Find the number.

   Equation: _____

   Answer: _____

   Proof: _____

   _____

6. Three times the difference of a number and 5 is 18. Find the number.

   Equation: _____

   Answer: _____

   Proof: _____

   _____

Equations and Inequalities: Lesson 1B

Practice

Translate each of these sentences into an equation. Solve the equation. To prove that your answer is correct, read the problem with the answer in place. If it makes a true statement, write *true*.

1. The difference between a number and 2 is 8. Find the number.

Equation: _____

Answer: _____

Proof: _____

_____

2. The product of −7 and a number is −56. Find the number.

Equation: _____

Answer: _____

Proof: _____

_____

3. The sum of −5 and a number is −10. Find the number.

Equation: _____

Answer: _____

Proof: _____

_____

4. The quotient of a number and −4 is 20. Find the number.

Equation: _____

Answer: _____

Proof: _____

_____

5. Six more than twice a number is −12. Find the number.

Equation: _____

Answer: _____

Proof: _____

_____

6. Negative four times the sum of a number and 3 is 16. Find the number.

Equation: _____

Answer: _____

Proof: _____

_____

Equations and Inequalities: Lesson 2A

Practice

This inequality has been solved. For each step, fill in the blank with the explanation for the process that was used to reach that step.

$$4x - 8 - x + 3 \geq 16$$

1. $3x - 5 \geq 16$ _____

2. $3x - 5 + 5 \geq 16 + 5$ _____

3. $3x \geq 21$ _____

4. $\frac{3x}{3} \geq \frac{21}{3}$ _____

5. $x \geq 7$ _____

6.  _____

Solve each inequality and graph the solution. Watch out for negative coefficients.

7. $2x - 3 < 11$

Solution is _____.

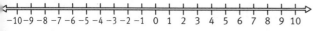

8. $-4(x + 3) + 5 > 1$

Solution is _____.

9. $3x - 5 \geq -26$

Solution is _____.

10. $2(3x + 1) - 5(x - 2) \leq 9$

Solution is _____.

Practice

This inequality has been solved. For each step, fill in the blank with the explanation for the process that was used to reach that step.

$6(2x + 3) - 8x < 20$

1. $12x + 18 - 8x < 20$ _____

2. $4x + 18 < 20$ _____

3. $4x + 18 - 18 < 20 - 18$ _____

4. $4x < 2$ _____

5. $\frac{4x}{4} < \frac{2}{4}$ _____

6. $x < \frac{1}{2}$ _____

7.  _____

Solve each inequality and graph the solution. Watch out for negative coefficients.

8. $5x - 2 > -27$

Solution is _____.

10. $4x + 6 - 8x - 2 < 12$

Solution is _____.

9. $-3(2x + 1) + 8 \le -13$

Solution is _____.

Name _____ Date _____

Equations and Inequalities: Lesson 3A

Practice

Write and solve each word problem using an inequality, and then answer the question.

1. Meyer's Recycling Service has two pricing structures for its customers. Every container that is put out for pick up by the service must have a "paid" sticker on it. The fees for this service are outlined below:

   Plan A: $8 per month plus $0.25 per sticker

   Plan B: $5 per month plus $0.75 per sticker

   How many containers would a customer need to put out in a month's time for Plan A to be more cost effective than plan B?

Fill in the following table and then answer the question.

Steps	Problem
1) Identify the variable	
2) Express all unknown quantities in terms of the variable	
3) Set up the model	
4) Solve and conclude	

Write and solve each word problem using an inequality.

2. Four times the sum of a number and 5 is less than 48. Find the numbers that will satisfy this relationship.

   Inequality: _____

   Solution: _____

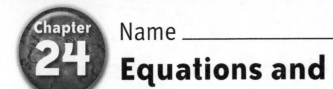
Practice

Write and solve each word problem using an inequality, and then answer the question.

1. Valley Electric Co–Op has two pricing structures for electric service for its rural customers. Each plan has a flat fee and then a rate for each kilowatt hour used.

   Plan A: $30 per month plus $0.04 per KWH (kilowatt hour)

   Plan B: $12 per month plus $0.06 per KWH

   How many kilowatt hours would a customer need to use in a month's time for Plan A to be more cost effective than plan B?

Fill in the following table and then answer the question.

Steps	Problem
1) Identify the variable	
2) Express all unknown quantities in terms of the variable	
3) Set up the model	
4) Solve and conclude	

Write and solve each word problem using an inequality.

2. The product of 6 and a number, decreased by 9, is less than or equal to 15. Find the numbers that will satisfy this relationship.

   Inequality: _____

   Solution: _____

Name _____ Date _____

Equations and Inequalities: Lesson 4A

Practice

Give an example of each of the following mathematical terms.

1. Exponent:

_____

4. Like terms:

_____

2. Variable:

_____

5. Distributive Property:

_____

3. Multiplication Property of Equality:

_____

6. Reciprocal:

_____

Solve the following word problem: Three times the sum of a number and 4 is equal to the difference between that number and 10. Find the number.

7. Equation: _____

8. Solution: _____

9. Proof: _____

_____

Solve the inequality $-3(x + 2) + 7x \leq 2x - 10$. Show your work and write the explanation for what was done to reach each step. Graph your solution on the number line.

10. Solution: _____

Graph: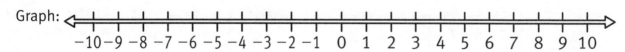

Equations and Inequalities: Lesson 4B

Practice

Give an example of each of the following mathematical terms.

1. Opposite:

_____

4. Unlike terms:

_____

2. Equation:

_____

5. Associative Property:

_____

3. Addition Property of Equality:

_____

6. Base:

_____

Solve the following word problem: Four times the difference between a number and 7 is equal to the sum of that number and 5. Find the number.

7. Equation: _____

8. Solution: _____

9. Proof: _____

_____

Solve the inequality $-2(x - 5) - 4 > -4x + 18$. Show your work and write the explanation for what was done to reach each step. Graph your solution on the number line.

10. Solution: _____

Graph:

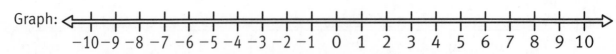

Chapter 1
Name _____ Date _____

Place Value and Addition: Lesson 1A

Practice

Write the greater number.

1. 2,361   2,163   __2,361__   3. 3,742   3,762   __3,762__

2. 1,385   1,359   __1,385__   4. 7,042   7,024   __7,042__

For each pair, write the place value where the numbers differ.

5. 6,243   6,043   7. 8,315   9,315
   __hundreds__   __thousands__

6. 7,214   7,274   8. 2,046   2,346
   __tens__   __hundreds__

Fill in the blank with a digit to make each sentence true.

9. 6,145 < 6,1_3   11. 3,429 > 3,_31
   __5, 6, 7, 8, or 9__   __0, 1, 2, or 3__

10. 5,432 > 5_57   12. 728 < _40
   __0, 1, 2, or 3__   __7, 8, or 9__

Chapter 1

Name _____ Date _____

Place Value and Addition: Lesson 1B

Practice

Write the greater number.

1. 1,823   1,832   __1,832__   3. 7,145   7,154   __7,154__

2. 6,827   6,817   __6,827__   4. 5,341   5,431   __5,431__

For each pair, write the place value where the numbers differ.

5. 8,142   8,162   7. 5,245   5,345
   __tens__   __hundreds__

6. 6,187   9,187   8. 4,216   4,219
   __thousands__   __ones__

Fill in the blank with a digit to make each sentence true.

9. 3,267 < 3,_41   10. 5,261 > 5,2_0
   __3, 4, 5, 6, 7, 8, or 9__   __0, 1, 2, 3, 4, 5, or 6__

Chapter 1

Name _____ Date _____

Place Value and Addition: Lesson 2A

Practice

For each pair, write the place value where the numbers differ.

1. 625   635   3. 3,215   3,915
   __tens__   __hundreds__

2. 4,034   5,034   4. 826   820
   __thousands__   __ones__

Write these numbers in order from least to greatest.

5. 643; 634; 642; 624; 613
   __613; 624; 634; 642; 643__

6. 384; 348; 483; 434; 388
   __348; 384; 388; 434; 483__

7. 673; 637; 763; 376; 676
   __376; 637; 673; 676; 763__

8. 4,216; 4,612; 4,618; 4,681; 4,482
   __4,216; 4,482; 4,612; 4,618; 4,681__

9. 1,035; 1,350; 1,503; 1,053; 1,353
   __1,035; 1,053; 1,350; 1,353; 1,503__

10. 8,634; 8,436; 8,346; 8,643; 8,364
   __8,346; 8,364; 8,436; 8,634; 8,643__

Chapter 1

Name _____ Date _____

Place Value and Addition: Lesson 2B

Practice

For each pair, write the place value where the numbers differ.

1. 8,024   8,224   3. 7,125   7,127
   __hundreds__   __ones__

2. 5,429   7,429   4. 6,352   6,392
   __thousands__   __tens__

Write these numbers in order from least to greatest.

5. 76; 67; 73; 63; 66
   __63; 66; 67; 73; 76__

6. 82; 208; 28; 288; 802
   __28; 82; 208; 288; 802__

7. 143; 134; 341; 314; 144
   __134; 143; 144; 314; 341__

8. 673; 376; 637; 367; 736
   __367; 376; 637; 673; 736__

9. 2,136; 2,316; 2,216; 2,326; 2,236
   __2,136; 2,216; 2,236; 2,316; 2,326__

10. 4,135; 4,105; 4,503; 4,351; 4,531
   __4,105; 4,135; 4,351; 4,503; 4,531__

Place Value and Addition: Lesson 3A

Practice

Find the sum. Write your answer.

1. 316 + 522 **838**	6. 138 + 831 **969**
2. 312 + 564 **876**	7. 625 + 312 **937**
3. 214 + 604 **818**	8. 451 + 328 **779**
4. 516 + 220 **736**	9. 443 + 216 **659**
5. 342 + 518 **860**	10. 625 + 204 **829**

Place Value and Addition: Lesson 3B

Practice

Find the sum. Write your answer.

1. 1237 + 5814 **7051**	6. 824 + 369 **1193**
2. 658 + 820 **1478**	7. 654 + 2351 **3005**
3. 4215 + 3824 **8039**	8. 8261 + 1359 **9620**
4. 4315 + 3266 **7581**	9. 4217 + 1058 **5275**
5. 743 + 8615 **9358**	10. 1625 + 2047 **3672**

Place Value and Addition: Lesson 4A

Practice

Find each sum. Write your answer. Write how many times you would make trades if you were using base-ten Blocks.

1. 256 + 145 **401** **2** trades	5. 518 + 298 **816** **2** trades
2. 342 + 521 **863** **0** trades	6. 358 + 824 **1182** **2** trades
3. 321 + 824 **1145** **1** trades	7. 634 + 270 **904** **1** trades
4. 424 + 618 **1042** **2** trades	8. 268 + 785 **1053** **3** trades

Place Value and Addition: Lesson 4B

Practice

Find each sum. Write your answer. Write how many times you would make trades if you were using Base-Ten Blocks.

1. 326 + 547 **873** **2** trades	5. 246 + 172 **418** **1** trades
2. 224 + 176 **400** **2** trades	6. 814 + 176 **990** **1** trades
3. 158 + 375 **533** **2** trades	7. 675 + 204 **879** **0** trades
4. 381 + 456 **837** **1** trades	8. 754 + 787 **1541** **3** trades

Place Value and Subtraction: Lesson 1A

Name _____ Date _____

Practice

Use the digits given to make the largest number possible using each digit one time. Write your answer.

1. 2, 4, 5 __542__   4. 2, 1, 6 __621__

2. 3, 5, 4 __543__   5. 7, 9, 2 __972__

3. 1, 6, 3 __631__   6. 3, 0, 8 __830__

Use the digits given to make the smallest number possible using each digit one time. Write your answer.

7. 3, 2, 6 __236__   10. 4, 3, 5 __345__

8. 6, 2, 7 __267__   11. 7, 1, 9 __179__

9. 2, 1, 4 __124__   12. 6, 3, 8 __368__

Algebra Readiness • Practice **17**

Place Value and Subtraction: Lesson 1B

Name _____ Date _____

Practice

Use the digits given to make the largest number possible using each digit one time. Write your answer.

1. 6, 7, 1 __761__   4. 0, 4, 2 __420__

2. 3, 9, 2 __932__   5. 6, 1, 8 __861__

3. 2, 3, 5 __532__   6. 5, 9, 3 __953__

Use the digits given to make the smallest number possible using each digit one time. Write your answer.

7. 4, 5, 2 __245__   10. 4, 2, 8 __248__

8. 3, 6, 2 __236__   11. 8, 2, 4 __248__

9. 3, 7, 1 __137__   12. 7, 3, 5 __357__

18 Algebra Readiness • Practice

Place Value and Subtraction: Lesson 2A

Name _____ Date _____

Practice

Choose the two numbers from each group that have the greatest difference. Write and solve that number sentence.

1. 326 541 822   4. 456 134 721
   __822 − 326 = 496__   __721 − 134 = 587__

2. 302 264 591   5. 312 545 748
   __591 − 264 = 327__   __748 − 312 = 436__

3. 143 624 250   6. 240 653 420
   __624 − 143 = 481__   __653 − 240 = 413__

Choose the two numbers from each group that have the smallest difference. Write and solve that number sentence.

7. 241 365 132   10. 164 358 571
   __241 − 132 = 109__   __358 − 164 = 194__

8. 326 453 271   11. 580 640 510
   __326 − 271 = 55__   __640 − 580 = 60__

9. 674 252 417   12. 670 320 840
   __417 − 252 = 165__   __840 − 670 = 170__

Algebra Readiness • Practice **19**

Place Value and Subtraction: Lesson 2B

Name _____ Date _____

Practice

Choose the two numbers from each group that have the greatest difference. Write and solve that number sentence.

1. 414 621 220   4. 351 220 515
   __621 − 220 = 401__   __515 − 220 = 295__

2. 246 648 420   5. 216 420 152
   __648 − 246 = 402__   __420 − 152 = 268__

3. 145 532 726   6. 412 752 146
   __726 − 145 = 581__   __752 − 146 = 606__

Choose the two numbers from each group that have the smallest difference. Write and solve that number sentence.

7. 134 625 415   10. 216 375 425
   __625 − 415 = 210__   __425 − 375 = 50__

8. 616 420 514   11. 515 721 312
   __514 − 420 = 94__   __515 − 312 = 203__

9. 630 428 534   12. 140 570 330
   __630 − 534 = 96__   __330 − 140 = 190__

20 Algebra Readiness • Practice

Name _____ **Date** _____

Place Value and Subtraction: Lesson 3A

Practice

Find each difference. Write your answer. Use addition to check your answer.

1.
```
   421
 -   8
   413
```
Check:
```
     8
 +413
   421
```

2.
```
   618
 -  73
   545
```
Check:
```
    73
 +545
   618
```

3.
```
   256
 -175
    81
```
Check:
```
   175
 +  81
   256
```

4.
```
   734
 -660
    74
```
Check:
```
   660
 +  74
   734
```

5.
```
   212
 -146
    66
```
Check:
```
   146
 +  66
   212
```

6.
```
   565
 -348
   217
```
Check:
```
   348
 +217
   565
```

Name _____ **Date** _____

Place Value and Subtraction: Lesson 3B

Practice

Find each difference. Write your answer. Use addition to check your answer.

1.
```
   624
 -   8
   616
```
Check:
```
     8
 +616
   624
```

2.
```
   152
 -  37
   115
```
Check:
```
    37
 +115
   152
```

3.
```
   537
 -  14
   523
```
Check:
```
    14
 +523
   537
```

4.
```
   843
 -  56
   787
```
Check:
```
    56
 +787
   843
```

5.
```
   715
 -  99
   616
```
Check:
```
    99
 +616
   715
```

6.
```
   653
 -142
   511
```
Check:
```
   142
 +511
   653
```

7.
```
   257
 -238
    19
```
Check:
```
   238
 +  19
   257
```

8.
```
   645
 -327
   318
```
Check:
```
   327
 +318
   645
```

Name _____ **Date** _____

Place Value and Subtraction: Lesson 4A

Practice

Write the missing digits in each subtraction problem.

1.
```
   328
 -215
   113
```

5.
```
   436
 -252
   184
```

2.
```
   754
 -371
   383
```

6.
```
   624
 -175
   449
```

3.
```
   176
 - 58
   118
```

7.
```
   335
 -148
   187
```

4.
```
   374
 -212
   162
```

8.
```
   581
 -208
   373
```

Name _____ **Date** _____

Place Value and Subtraction: Lesson 4B

Practice

Write the missing digits in each subtraction problem.

1.
```
   137
 - 24
   113
```

6.
```
   644
 -358
   286
```

2.
```
   242
 -153
    89
```

7.
```
   468
 -355
   113
```

3.
```
   427
 -156
   271
```

8.
```
   379
 -168
   211
```

4.
```
   345
 -183
   162
```

9.
```
   627
 -445
   182
```

5.
```
   324
 -161
   163
```

10.
```
   451
 -166
   285
```

204   Algebra Readiness • Practice Answers

Name _____ Date _____
Multiplication: Lesson 1A

Practice

Write the multiplication sentence modeled by each array.

1.

   _4_ × _5_ = _20_

4.

   _2_ × _7_ = _14_

2.

   _3_ × _4_ = _12_

5.

   _3_ × _7_ = _21_

3.

   _8_ × _2_ = _16_

6.

   _8_ × _3_ = _24_

Name _____ Date _____
Multiplication: Lesson 1B

Practice

Write the multiplication sentence modeled by each array.

1.

   _5_ × _8_ = _40_

4.

   _2_ × _10_ = _20_

2.

   _3_ × _11_ = _33_

5.

   _7_ × _4_ = _28_

3.

   _9_ × _2_ = _18_

6.

   _5_ × _9_ = _45_

Name _____ Date _____
Multiplication: Lesson 2A

Practice

Fill in the blanks to find each product.

1. 5 × 72

   (_5_ × _70_) + (_5_ × _2_)

   _350_ + _10_

   _360_

3. 14 × 63

   (_14_ × _60_) + (_14_ × _3_)

   _840_ + _42_

   _882_

2. 8 × 36

   (_8_ × _30_) + (_8_ × _6_)

   _240_ + _48_

   _288_

4. 23 × 64

   (_23_ × _60_) + (_23_ × _4_)

   _1,380_ + _92_

   _1,472_

Find each product. Write your answer.

5.
   38
   × 5
   ___
   190

8.
   742
   × 5
   ___
   3,710

6.
   59
   × 6
   ___
   354

9.
   47
   × 3
   ___
   141

7.
   604
   × 9
   ___
   5,436

10.
   82
   × 31
   ___
   2,542

Name _____ Date _____
Multiplication: Lesson 2B

Practice

Fill in the blanks to find each product.

1. 6 × 54

   (_6_ × _50_) + (_6_ × _4_)

   _300_ + _24_

   _324_

3. 17 × 32

   (_17_ × _30_) + (_17_ × _2_)

   _510_ + _34_

   _544_

2. 4 × 67

   (_4_ × _60_) + (_4_ × _7_)

   _240_ + _28_

   _268_

4. 34 × 72

   (_34_ × _70_) + (_34_ × _2_)

   _2,380_ + _68_

   _2,448_

Find each product. Write your answer.

5.
   43
   × 9
   ___
   387

8.
   408
   × 7
   ___
   2,856

6.
   27
   × 4
   ___
   108

9.
   63
   × 8
   ___
   504

7.
   507
   × 6
   ___
   3,042

10.
   126
   × 34
   ___
   4,284

Algebra Readiness • Practice Answers 205

Chapter 3

Name _____ Date _____

Multiplication: Lesson 3A

Practice

Use the partial-product method to find each product. Show your work.
Write your answer.

1.
```
    54
 ×   7
  350
 + 28
  378
```

4.
```
   342
 ×   8
  2400
   320
 +  16
  2736
```

2.
```
    45
 ×   6
  240
 + 30
  270
```

5.
```
   245
 ×  36
  6000
  1200
   150
  1200
   240
 +  30
  8820
```

3.
```
   426
 ×   3
  1200
    60
 +  18
  1278
```

6.
```
   245
 ×  42
  8000
  1600
   200
   400
    80
 +  10
 10290
```

Chapter 3

Name _____ Date _____

Multiplication: Lesson 3B

Practice

Use the partial-product method to find each product. Show your work.
Write your answer.

1.
```
    37
 ×   4
  120
 + 28
  148
```

4.
```
   615
 ×   7
  4200
    70
 +  35
  4305
```

2.
```
    59
 ×   4
  200
 + 36
  236
```

5.
```
   321
 ×  27
  6000
   400
    20
  2100
   140
 +   7
  8667
```

3.
```
   519
 ×   6
  3000
    60
 +  54
  3114
```

6.
```
   314
 ×  53
 15000
   500
   200
   900
    30
 +  12
 16642
```

Chapter 3

Name _____ Date _____

Multiplication: Lesson 4A

Practice

Solve each multiplication problem. Write your answer. Then match the
problem with the description of a grid that is large enough to hold the
problem's array with the fewest number of squares left over. Use graph
paper if you need help visualizing the grids.

1. 3 × 8 = __24__

2. 12 × 6 = __72__

3. 7 × 21 = __147__

4. 16 × 5 = __80__

5. 6 × 18 = __108__

6. 23 × 8 = __184__

A. a grid with 8 columns and 22 rows

B. a grid with 18 columns and 8 rows

C. a grid with 4 columns and 10 rows

D. a grid with 24 columns and 10 rows

E. a grid with 14 columns and 7 rows

F. a grid with 16 columns and 6 rows

Chapter 3

Name _____ Date _____

Multiplication: Lesson 4B

Practice

Solve each multiplication problem. Write your answer. Then match the
problem with the description of a grid that is large enough to hold the
problem's array with the fewest number of squares left over. Use graph
paper if you need help visualizing the grids.

1. 5 × 12 = __60__

2. 7 × 3 = __21__

3. 6 × 20 = __120__

4. 3 × 16 = __48__

5. 22 × 9 = __198__

6. 17 × 6 = __102__

A. a grid with 8 columns and 5 rows

B. a grid with 4 columns and 18 rows

C. a grid with 5 columns and 15 rows

D. a grid with 18 columns and 7 rows

E. a grid with 7 columns and 20 rows

F. a grid with 24 columns and 10 rows

---

206   Algebra Readiness • Practice Answers

Division: Lesson 1A

Practice

Solve each problem below using counters or drawing circles. Write a
division number sentence to describe each situation.

1. Kacy is setting up the school dining room for a special luncheon. There
   are 6 tables in the room. If 42 people are coming to the luncheon, how
   many places should Kacy set at each table so the same number of
   guests are seated at each table?

   ___7___ guests at each table   42 ÷ 6 = 7

2. The sixth graders are organizing into 4 teams to play soccer. If there are
   48 sixth graders, how many will be on each team?

   ___12___ sixth graders on each team   48 ÷ 4 = 12

3. One hundred fifty-five students are going on a field trip to the art
   museum. Five buses will be used to transport the students. How
   many students will be on each bus?

   ___31___ students will ride on each bus.   155 ÷ 5 = 31

4. Juanita is planting 60 tulip bulbs in 4 flower beds in her backyard. She
   wants to put the same number of bulbs in each bed. How many bulbs
   should be planted in each flower bed?

   ___15___ bulbs in each bed   60 ÷ 4 = 15

Division: Lesson 1B

Practice

Solve each problem below using Counters or drawing circles. Write a
division number sentence to model each situation.

1. Forty students are organized into 8 teams for basketball drills. How many
   students will be on each team?

   ___5___ students on each team   40 ÷ 8 = 5

2. In the last 3 days, the high temperature increased a total of 18 degrees.
   The increase in temperature was the same for each day. How many
   degrees did the temperature increase each day?

   ___6___ degrees each day   18 ÷ 3 = 6

3. Forty-eight books need to be mailed in 6 cartons. The same
   number of books needs to be placed in each carton. How many books
   should be put into each box?

   ___8___ books in each carton   48 ÷ 6 = 8

4. José jogged a total of 28 miles in 4 days. Since he was training, he
   jogged the same number of miles each day. How many miles did he
   jog each day?

   ___7___ miles each day   28 ÷ 4 = 7

5. Sixteen chemistry students are put on 4 teams to work on a science
   fair project. How many students will be on each team?

   ___4___ students on each team   16 ÷ 4 = 4

6. Aretha has 12 thank-you notes to write for birthday presents she
   received. She wants to write the same number of notes each day for
   the next 4 days. How many notes should she write each day?

   ___3___ notes each day   12 ÷ 4 = 3

Division: Lesson 2A

Practice

Solve each problem below using Counters or drawing circles. Write a
division number sentence to model each situation.

1. Flu vaccine comes in 30 cubic-centimeter vials. Each dose is 4 cubic centimeters.
   How many doses can be given from each vial?

   7 doses of vaccine; 2 cc left

   30 ÷ 4 = 7 R 2

2. Eldora has 20 daisy plants to put in the 3 flower beds in front of her
   house. She wants each bed to be the same, and she'll put any extra
   plants in her backyard. How many plants should she put in each bed
   in the front yard?

   6 plants in each bed in the front yard; 2 plants left

   20 ÷ 3 = 6 R 2

3. There are 23 fifth graders in Mr. Ong's physical education class. He
   organizes the students into 4 even teams. The remaining students act
   as referees. How many students will be on each team?

   5 students on each team; 3 left to be referees

   23 ÷ 4 = 5 R 3

4. The hospitality committee at the community center is making fall gift
   baskets for 7 people. The committee has 30 apples to put in the
   baskets. How many apples should be put in each basket?

   4 apples in each basket; 2 apples left

   30 ÷ 7 = 4 R 2

Division: Lesson 2B

Practice

Solve each problem below using Counters or drawing circles. Write a
division number sentence to model each situation.

1. In preparation for the next 10-kilometer race, Marty wants to run a total of
   26 miles this week. He will run the same distance on Monday,
   Wednesday, and Thursday, and finish the remainder of the miles on
   Friday. How many miles should he run on each of the first three days?

   8 miles on each of the first three days

   2 miles left; 26 ÷ 3 = 8 R 2

2. There are 5 display rooms in the local art gallery. The new exhibit has
   a total of 48 pieces of art. The curator will display the same number of
   pieces of art in each display room, and display the remainder of the pieces
   in the lobby. How many pieces of art will be in each display room?

   9 pieces of art in each room

   3 leftover pieces in the lobby; 48 ÷ 5 = 9 R 3

3. The fourth graders are getting ready for a softball game. There are
   27 students in the fourth grade class and they will be organized evenly
   into 2 teams. The remaining student will be the third-base coach. How
   many students will be on each team?

   13 students on each team

   1 student coach 27 ÷ 2 = 13 R 1

4. The school cafeteria staff is making up box lunches for a fifth grade
   picnic. They have 50 cookies to put into the box lunches. Each box is
   to get 3 cookies. How many boxes can they fill with the cookies?

   16 box lunches; 2 cookies left

   50 ÷ 3 = 16 R 2

Chapter 4

Name _____ **Date** _____

Division: Lesson 3A

Practice

Solve each problem. Show your work. Write your answer.

1. Ms. Sanchez's students are coming to her classroom for the first day. The classroom has 8 rows of desks with 4 desks in each row. There are 30 students enrolled in her class. How many full rows of students will there be? How many students will sit in the last row?

   __7__ full rows of students with __2__ student(s) in the last row

2. Mr. Perez is fixing fruit for his 3 children to take in their school lunches. He has 8 tangerines. How many tangerines can he put in each child's lunch? How many tangerines will he have left?

   __2__ tangerines in each lunch with __2__ tangerine(s) left

3. Marlena is making costumes for her little nieces and nephews for trick-or-treat night. She has 11 yards of shiny silver cotton fabric, and each costume requires 2 yards of fabric. How many costumes can she make? How many yards of fabric will she have left?

   __5__ costumes with __1__ yard(s) of fabric left

4. Tommy, a worker in the produce department at the grocery store, is putting together bunches of green onions. He'll put 8 onions in each bunch. The delivery this morning includes 260 green onions. How many bunches can he assemble? How many green onions will be left?

   __32__ bunches of green onions with __4__ green onion(s) left

---

Chapter 4

Name _____ **Date** _____

Division: Lesson 3B

Practice

Solve each problem. Show your work. Write your answers.

1. Tammy is organizing her CD library. She has 98 CDs to put on shelves that will each hold 8 CDs. How many full shelves will she have? How many CDs will she have left?

   __12__ full shelves of CDs with __2__ CD(s) left

2. Absha is making practice skirts for the students in her dance class. Each skirt requires 2 yards of fabric. She has 25 yards of pink net material to use to make the skirts. How many skirts can she make? How many yards of material will be left?

   __12__ practice skirts with __1__ yard(s) left

3. The service club at the local high school is gathering canned goods to send to victims of the recent earthquake. Each carton will hold 8 large cans of juice. How many cartons will they need to pack to hold the 75 large cans of juice they have collected? How many large cans of juice will be left?

   __9__ cartons of large juice cans with __3__ can(s) of juice left

4. Don's Roofing Service has received a shipment of 25 rolls of roofing paper to use for their next few jobs. Each job requires 4 rolls of roofing paper. How many jobs can they complete with this roofing paper? How many rolls will be left?

   __6__ roofing jobs with __1__ roll(s) left

5. In Tony's coin collection there are 118 aluminum cents. These cents can be put into plastic holders that will hold 6 cents each. How many plastic holders will he need for his groups of Indian Head cents? How many cents will be left?

   __19__ plastic holders with __4__ cent(s) left

6. Xander is putting his baseball cards into stacks of 8 cards each so that he can make trades with his friends. He has 55 cards that he wants to trade. How many stacks can he make? How many cards will be left?

   __6__ stacks of cards with __7__ card(s) left

Chapter 4

Name _____ **Date** _____

Division: Lesson 4A

Practice

Find the remainder in each situation. Then determine whether it makes sense to round up, round down, or divide the remainder evenly. Explain. Write your answer.

1. To play a certain card game, you take the cards in the deck (there are 52 cards in a deck) and pass them out evenly to all 6 players. The cards that are left you place the facedown in the "discard" pile.

   The 52 cards can be divided into 6 groups of 8 with 4 cards left. It makes sense to round down since you don't have enough cards for another player.

2. Each player will have __8__ cards.

3. Carolyn uses 12 quilt squares to make a baby quilt. She has 50 quilt squares ready to use when the next baby is born.

   The 50 quilt squares can be divided into 4 groups of 12 squares each, and there will be 2 quilt squares left. It makes sense to round down because there are not enough squares for another quilt.

4. Carolyn can make __4__ baby quilts with the squares she has.

5. A local farmer has brought 6 dozen apples to the primary school to be used for snacks for the children. There are 4 classrooms in the primary school.

   The 6 dozen apples can be divided by 4, and each classroom will get 1 dozen apples. It makes sense to divide the remaining 2 dozen apples evenly between the classrooms.

6. Each classroom can have __$1\frac{1}{2}$__ dozen apples.

Chapter 4

Name _____ **Date** _____

Division: Lesson 4B

Practice

Find the remainder in each situation. Then determine whether it makes sense to round up, round down, or divide the remainder evenly. Explain. Write your answer.

1. The band is taking school vans to this week's football game. Each van can hold 9 students. There are 98 students in the band.

   The 98 students can be divided in groups of 9. They will need 10 vans. It makes sense to round up because the 8 remaining students also need to go to the game too.

2. The band will need __11__ vans.

3. Clarissa uses 8 triangular pieces of cloth to make a placemat. She has 35 triangular pieces cut and ready to use.

   The 35 pieces can be separated into 4 stacks of 8 each and there will be 3 triangular pieces left over. It makes sense to round down because the 3 pieces aren't enough to finish another placemat.

4. Clarissa can make __4__ placemats from the pieces she has.

5. Don has $63 in his change jar to give to his 6 grandchildren.

   The $63 can be separated into 6 stacks of change worth $10 each, and there will be $3 left. It makes sense to split the remaining $3 between the 6 grandchildren.

6. Each child will receive __$10 and $\frac{1}{2}$ dollars__

Parts of a Whole: Lesson 1A

Practice

Write each fraction modeled.

1.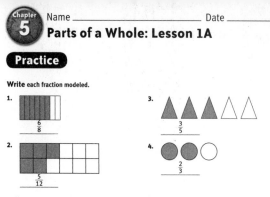
$\frac{6}{8}$

2. $\frac{5}{12}$

3. $\frac{3}{5}$

4. $\frac{2}{3}$

Follow each instruction.

5. Divide the whole set below into 3 parts by circling individual sets.

6. Now shade $\frac{2}{3}$ of the squares.
Students should shade 10 of the 15 squares.

7. Below is $\frac{3}{4}$ of a set of pentagons. Draw the number of pentagons needed for a complete set.

8. Below is $\frac{1}{5}$ of a set of squares. Draw the number of squares needed for a complete set.

Parts of a Whole: Lesson 1B

Practice

Write each fraction shown.

1. $\frac{3}{4}$

2. $\frac{6}{10}$

3. $\frac{2}{6}$

4. $\frac{1}{4}$

Follow each instruction.

5. Divide the whole set below into 4 parts by circling individual sets.

6. Now shade $\frac{3}{4}$ of the squares.
Students should shade 15 of the 20 squares.

7. Below is $\frac{2}{5}$ of a set of circles. Draw the number of circles needed for a complete set.

8. Below is $\frac{2}{6}$ of a set of triangles. Draw the number of triangles needed for a complete set.

Parts of a Whole: Lesson 2A

Practice

Divide each shape into the number of equal parts indicated by the denominator. Then shade the unit fraction as indicated by the numerator. Write the fractional amount that is not shaded.
Sample answers shown.

1. $\frac{3}{4}$
$\frac{1}{4}$ is not shaded.

2. $\frac{4}{6}$
$\frac{2}{6}$ is not shaded.

3. $\frac{3}{8}$
$\frac{5}{8}$ is not shaded.

4. $\frac{7}{10}$
$\frac{3}{10}$ is not shaded.

Draw each whole set.

5. Shown is $\frac{2}{7}$ of a set.

6. Shown is $\frac{3}{6}$ of a set.

Parts of a Whole: Lesson 2B

Practice

Divide each shape into the number of equal parts indicated by the denominator. Then shade the unit fraction as indicated by the numerator. Write the fractional amount that is not shaded.
Sample answers shown.

1. $\frac{2}{5}$
$\frac{3}{5}$ is not shaded.

2. $\frac{2}{3}$
$\frac{1}{3}$ is not shaded.

3. $\frac{1}{6}$
$\frac{5}{6}$ is not shaded.

4. $\frac{5}{8}$
$\frac{3}{8}$ is not shaded.

Draw each whole set.

5. Shown is $\frac{2}{5}$ of a set.

6. Shown is $\frac{4}{9}$ of a set.

Algebra Readiness • Practice Answers   209

Name _____ Date _____

Parts of a Whole: Lesson 3A

Practice

Make one whole using the same unit fraction. Write how many more Fraction Bars you will need to make 1.

1.

| 1 |

| $\frac{1}{4}$ | $\frac{1}{4}$ | $\frac{1}{4}$ |

___1___ more $\frac{1}{4}$ Fraction Bars

2.

| 1 |

| $\frac{1}{8}$ | $\frac{1}{8}$ | $\frac{1}{8}$ |

___5___ more $\frac{1}{8}$ Fraction Bars

3.

| 1 |

| $\frac{1}{6}$ | $\frac{1}{6}$ |

___4___ more $\frac{1}{6}$ Fraction Bars

Draw a grid for each of the following fractions.

4. $\frac{8}{8}$
any rectangular configuration of 8 squares with all 8 squares shaded

5. $\frac{5}{5}$
any rectangular configuration of 5 squares with all 5 squares shaded

---

Name _____ Date _____

Parts of a Whole: Lesson 3B

Practice

Make one whole using the same unit fraction. Write how many more Fraction Bars you will need to make 1.

1.

| 1 |

| $\frac{1}{3}$ | $\frac{1}{3}$ |

___1___ more $\frac{1}{3}$ Fraction Bars

2.

| 1 |

| $\frac{1}{10}$ | $\frac{1}{10}$ | $\frac{1}{10}$ |

___7___ more $\frac{1}{10}$ Fraction Bars

3.

| 1 |

| $\frac{1}{8}$ | $\frac{1}{8}$ | $\frac{1}{8}$ | $\frac{1}{8}$ | $\frac{1}{8}$ | $\frac{1}{8}$ |

___2___ more $\frac{1}{8}$ Fraction Bars

Draw a grid for each of the following fractions.

4. $\frac{4}{4}$
any rectangular configuration of 4 squares with all 4 squares shaded

5. $\frac{12}{12}$
any rectangular configuration of 12 squares with all 12 squares shaded

---

Name _____ Date _____

Parts of a Whole: Lesson 4A

Practice

Write each total using both cent notation and dollar notation.

1.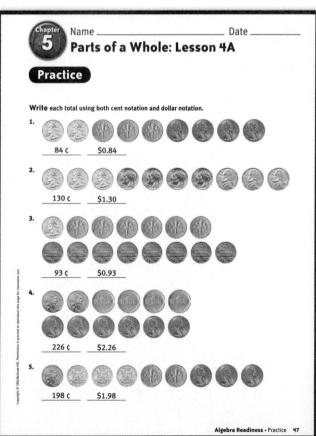
84 ¢ $0.84

2.
130 ¢ $1.30

3.
93 ¢ $0.93

4.
226 ¢ $2.26

5.
198 ¢ $1.98

Name _____ Date _____

Parts of a Whole: Lesson 4B

Practice

Write each total using both cent notation and dollar notation.

1.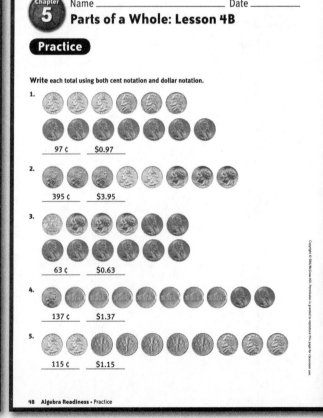
97 ¢ $0.97

2.
395 ¢ $3.95

3.
63 ¢ $0.63

4.
137 ¢ $1.37

5.
115 ¢ $1.15

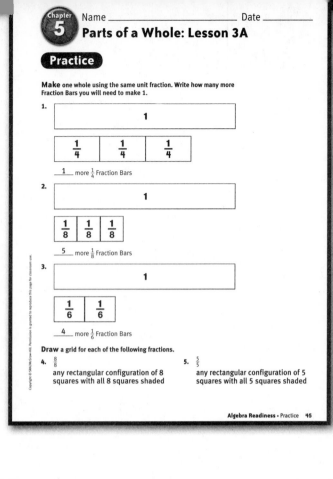
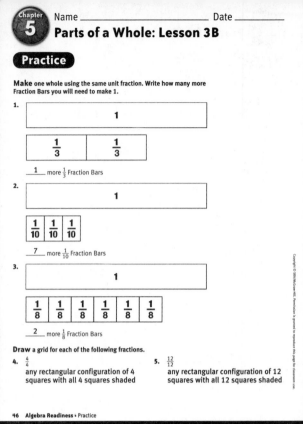

210   Algebra Readiness • Practice Answers

Name _____ Date _____

Positive and Negative Fractions: Lesson 1A

Practice

Do the fraction and the decimal match? Write *yes* or *no*.

1. $\frac{4}{5}$ and 0.5 ____no____
2. $\frac{3}{8}$ and 0.375 ____yes____
3. $\frac{1}{4}$ and 0.25 ____yes____
4. $\frac{2}{3}$ and 0.6666... ____yes____
5. $\frac{4}{10}$ and 0.04 ____no____
6. $\frac{9}{10}$ and 0.09 ____no____
7. $\frac{15}{100}$ and 0.015 ____no____
8. $\frac{7}{7}$ and 1.0 ____yes____

Complete the table.

	Mixed Number	Improper Fraction	Decimal
9.	$1\frac{1}{5}$	$\frac{6}{5}$	1.2
10.	$2\frac{2}{5}$	$\frac{12}{5}$	2.4
11.	$1\frac{1}{3}$	$\frac{4}{3}$	1.3333...
12.	$1\frac{3}{5}$	$\frac{8}{5}$	1.6
13.	$2\frac{3}{4}$	$\frac{11}{4}$	2.75
14.	$1\frac{1}{8}$	$\frac{9}{8}$	1.125
15.	$3\frac{1}{4}$	$\frac{13}{4}$	3.25
16.	$1\frac{2}{3}$	$\frac{5}{3}$	1.6666...

Name _____ Date _____

Positive and Negative Fractions: Lesson 1B

Practice

Do the fraction and the decimal match? Write *yes* or *no*.

1. $\frac{4}{8}$ and 0.5 ____yes____
2. $\frac{1}{8}$ and 0.8 ____no____
3. $\frac{1}{3}$ and 0.33 ____no____
4. $\frac{7}{10}$ and 0.7777... ____no____
5. $\frac{7}{10}$ and 0.07 ____no____
6. $\frac{9}{10}$ and 0.9 ____yes____
7. $\frac{21}{100}$ and 0.21 ____yes____
8. $\frac{3}{3}$ and 1.0 ____yes____

Complete the table.

	Mixed Number	Improper Fraction	Decimal
9.	$1\frac{2}{5}$	$\frac{7}{5}$	1.4
10.	$2\frac{2}{3}$	$\frac{8}{3}$	2.6666...
11.	$1\frac{1}{4}$	$\frac{5}{4}$	1.25
12.	$1\frac{4}{5}$	$\frac{9}{5}$	1.8
13.	$2\frac{3}{8}$	$\frac{19}{8}$	2.375
14.	$1\frac{1}{2}$	$\frac{3}{2}$	1.5
15.	$2\frac{3}{4}$	$\frac{11}{4}$	2.75
16.	$3\frac{2}{3}$	$\frac{11}{3}$	3.6666...

Name _____ Date _____

Positive and Negative Fractions: Lesson 2A

Practice

Match each fraction with its decimal form. Draw a line from the fraction to the decimal.

1. $\frac{-3}{4}$ G
2. $\frac{5}{8}$ J
3. $\frac{17}{100}$ I
4. $\frac{2}{5}$ F
5. $\frac{-5}{8}$ C
6. $\frac{-1}{5}$ E
7. $\frac{-7}{10}$ B
8. $\frac{3}{8}$ H
9. $\frac{-1}{4}$ D
10. $\frac{3}{1}$ A

A. 3
B. −0.7
C. −0.625
D. −0.25
E. −0.2
F. 0.4
G. −0.75
H. 0.375
I. 0.17
J. 0.625

Name _____ Date _____

Positive and Negative Fractions: Lesson 2B

Practice

Match each fraction with its decimal form. Draw a line from the fraction to the decimal.

1. $\frac{-2}{3}$ C
2. $\frac{-3}{8}$ F
3. $\frac{-3}{5}$ I
4. $\frac{5}{1}$ H
5. $\frac{-5}{8}$ A
6. $\frac{-23}{100}$ J
7. $\frac{-9}{10}$ B
8. $\frac{3}{8}$ E
9. $\frac{1}{4}$ D
10. $\frac{5}{8}$ G

A. −0.625
B. −0.9
C. −0.6666...
D. 0.25
E. 0.375
F. −0.375
G. 0.625
H. 5
I. −0.6
J. −0.23

Positive and Negative Fractions: Lesson 3A

Practice

Estimate where each fraction is on the number line. Label each with a point and the fraction.

1. $\frac{1}{4}$ and $\frac{2}{3}$

0   $\frac{1}{4}$ $\frac{1}{2}$ $\frac{2}{3}$ 1

3. $\frac{4}{10}$ and $\frac{1}{3}$

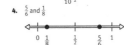

0   $\frac{4}{10}$ $\frac{1}{2}$ 1

2. $\frac{2}{5}$ and $\frac{5}{8}$

0   $\frac{2}{5}$ $\frac{5}{8}$ 1

4. $\frac{5}{6}$ and $\frac{1}{8}$

0   $\frac{1}{8}$ $\frac{1}{2}$ $\frac{5}{6}$ 1

Name a decimal between the given decimals. Record it on the number line.

5.

1.5   2.5

any number between 1.5 and 2.5

6.

0.4   1.0

any number between 0.4 and 1.0

7.

0.25   0.5

any number between 0.25 and 0.5

8.

0.6   0.7

any number between 0.6 and 0.7

Positive and Negative Fractions: Lesson 3B

Practice

Estimate where each fraction is on the number line. Label each with a point and the fraction.

1. $\frac{1}{5}$ and $\frac{5}{8}$

0   $\frac{1}{5}$ $\frac{1}{2}$ $\frac{5}{8}$ 1

3. $\frac{9}{10}$ and $\frac{5}{12}$

0   $\frac{5}{12}$$\frac{1}{2}$ $\frac{9}{10}$ 1

2. $\frac{1}{8}$ and $\frac{3}{4}$

0   $\frac{1}{8}$ $\frac{1}{2}$ $\frac{3}{4}$ 1

4. $\frac{3}{5}$ and $\frac{1}{3}$

0   $\frac{1}{3}$ $\frac{1}{2}$ $\frac{3}{5}$ 1

Name a decimal between the given decimals. Record it on the number line.

5.

0.25   0.75

any number between 0.25 and 0.75

6.

0.3   1.0

any number between 0.3 and 1.0

7.

0.15   0.45

any number between 0.15 and 0.45

8.

0.3   0.4

any number between 0.3 and 0.4

Positive and Negative Fractions: Lesson 4A

Practice

Estimate where each decimal is on the number line. Label each with a point and the decimal.

1. 0.4 and 0.8

0   0.4 $\frac{1}{2}$ 0.8   1

3. −0.6 and −0.1

−0.6   −0.1

−1 $-\frac{1}{2}$ 0

2. 0.125 and 0.75

0.125

0   $\frac{1}{2}$ 0.75   1

4. −0.25 and −0.8

−1 −0.8 $-\frac{1}{2}$ −0.25   0

Name a decimal between the given decimals. Record it on the number line.

5.

0.4   0.8

any number between 0.4 and 0.8

6.

0.5   0.6

any number between 0.5 and 0.6

7.

−0.4   0.4

any number between −0.4 and 0.4

8.

−0.7   −0.65

any number between −0.7 and −0.65

Positive and Negative Fractions: Lesson 4B

Practice

Estimate where each decimal is on the number line. Label each with a point and the decimal.

1. 0.3 and 0.75

0   0.3 $\frac{1}{2}$ 0.75   1

3. −0.6 and −0.4

−0.6 −0.4

−1 $-\frac{1}{2}$ 0

2. 0.4 and 0.9

0.4   0.9

0   $\frac{1}{2}$ 1

4. −0.1 and −0.75

−0.75   −0.1

−1 $-\frac{1}{2}$ 0

Name a decimal between the given decimals. Record it on the number line.

5.

0.35   0.6

any number between 0.35 and 0.6

6.

0.2   0.3

any number between 0.2 and 0.3

7.

−0.25   0.25

any number between −0.25 and 0.25

8.

−0.95   −0.8

any number between −0.95 and −0.8

Name _____ Date _____

Prime Factorization and Powers of 10: Lesson 1A

Practice

Write the factors in each expression.

1. $3 \times 9 = 27$ ___3___ ___9___

2. $11 \times 6 = 66$ ___11___ ___6___

3. $5 \times 12 = 60$ ___5___ ___12___

4. $8 \times 15 = 120$ ___8___ ___15___

Write the first five multiples of each number.

5. 3 ___3, 6, 9, 12, 15___

6. 12 ___12, 24, 36, 48, 60___

7. 8 ___8, 16, 24, 32, 40___

8. 7 ___7, 14, 21, 28, 35___

Write whether the following numbers are *prime* or *composite*.

9. 10 is a ___composite___ number.

10. 5 is a ___prime___ number.

11. 17 is a ___prime___ number.

12. 35 is a ___composite___ number.

13. 27 is a ___composite___ number.

14. 43 is a ___prime___ number.

15. 9 is a ___composite___ number.

Name _____ Date _____

Prime Factorization and Powers of 10: Lesson 1B

Practice

Write the factors in each expression.

1. $6 \times 4 = 24$ ___6___ ___4___

2. $14 \times 3 = 42$ ___14___ ___3___

3. $8 \times 10 = 80$ ___8___ ___10___

4. $7 \times 12 = 84$ ___7___ ___12___

Write the first five multiples of each number.

5. 6 ___6, 12, 18, 24, 30___

6. 9 ___9, 18, 27, 36, 45___

7. 11 ___11, 22, 33, 44, 55___

8. 15 ___15, 30, 45, 60, 75___

Write whether the following numbers are *prime* or *composite*.

9. 15 is a ___composite___ number.

10. 7 is a ___prime___ number.

11. 23 is a ___prime___ number.

12. 6 is a ___composite___ number.

13. 4 is a ___composite___ number.

14. 41 is a ___prime___ number.

15. 13 is a ___prime___ number.

Name _____ Date _____

Prime Factorization and Powers of 10: Lesson 2A

Practice

Write the following numbers as the product of their prime factors. Use counters, factor trees, mental math, or paper and pencil to determine the prime factors for each number.

1. 20 ___$20 = 2 \times 2 \times 5$___

2. 45 ___$45 = 3 \times 3 \times 5$___

3. 54 ___$54 = 2 \times 3 \times 3 \times 3$___

4. 18 ___$18 = 2 \times 3 \times 3$___

5. 15 ___$15 = 3 \times 5$___

6. 84 ___$84 = 2 \times 2 \times 3 \times 7$___

7. 52 ___$52 = 2 \times 2 \times 13$___

8. 100 ___$100 = 2 \times 2 \times 5 \times 5$___

9. 28 ___$28 = 2 \times 2 \times 7$___

10. 26 ___$26 = 2 \times 13$___

Name _____ Date _____

Prime Factorization and Powers of 10: Lesson 2B

Practice

Write the following numbers as the product of their prime factors. Use counters, factor trees, mental math, or paper and pencil to determine the prime factors for each number.

1. 30 ___$30 = 2 \times 3 \times 5$___

2. 55 ___$55 = 5 \times 11$___

3. 24 ___$24 = 2 \times 2 \times 2 \times 3$___

4. 28 ___$28 = 2 \times 2 \times 7$___

5. 35 ___$35 = 5 \times 7$___

6. 32 ___$32 = 2 \times 2 \times 2 \times 2 \times 2$___

7. 72 ___$72 = 2 \times 2 \times 2 \times 3 \times 3$___

8. 90 ___$90 = 2 \times 3 \times 3 \times 5$___

9. 17 ___$17 = 17$___

10. 42 ___$42 = 2 \times 3 \times 7$___

Practice

Write the following numbers as the product of their prime factors in expanded form and using exponents.

1. $8 = $ _____ $2 \times 2 \times 2$ _____ $= $ _____ 2^3 _____

2. $12 = $ _____ $2 \times 2 \times 3$ _____ $= $ _____ $2^2 \times 3$ _____

3. $54 = $ _____ $2 \times 3 \times 3 \times 3$ _____ $= $ _____ 2×3^3 _____

4. $120 = $ _____ $2 \times 2 \times 2 \times 3 \times 5$ _____ $= $ _____ $2^3 \times 3 \times 5$ _____

5. $35 = $ _____ 5×7 _____ $= $ _____ 5×7 _____

6. $32 = $ _____ $2 \times 2 \times 2 \times 2 \times 2$ _____ $= $ _____ 2^5 _____

7. $72 = $ _____ $2 \times 2 \times 2 \times 3 \times 3$ _____ $= $ _____ $2^3 \times 3^2$ _____

8. $80 = $ _____ $2 \times 2 \times 2 \times 2 \times 5$ _____ $= $ _____ $2^4 \times 5$ _____

9. $44 = $ _____ $2 \times 2 \times 11$ _____ $= $ _____ $2^2 \times 11$ _____

10. $56 = $ _____ $2 \times 2 \times 2 \times 7$ _____ $= $ _____ $2^3 \times 7$ _____

Practice

Write the following numbers as the product of their prime factors in expanded form and using exponents.

1. $16 = $ _____ $2 \times 2 \times 2 \times 2$ _____ $= $ _____ 2^4 _____

2. $24 = $ _____ $2 \times 2 \times 2 \times 3$ _____ $= $ _____ $2^3 \times 3$ _____

3. $77 = $ _____ 7×11 _____ $= $ _____ 7×11 _____

4. $80 = $ _____ $2 \times 2 \times 2 \times 2 \times 5$ _____ $= $ _____ $2^4 \times 5$ _____

5. $45 = $ _____ $3 \times 3 \times 5$ _____ $= $ _____ $3^2 \times 5$ _____

6. $100 = $ _____ $2 \times 2 \times 5 \times 5$ _____ $= $ _____ $2^2 \times 5^2$ _____

7. $63 = $ _____ $3 \times 3 \times 7$ _____ $= $ _____ $3^2 \times 7$ _____

8. $52 = $ _____ $2 \times 2 \times 13$ _____ $= $ _____ $2^2 \times 13$ _____

9. $34 = $ _____ 2×17 _____ $= $ _____ 2×17 _____

10. $150 = $ _____ $2 \times 3 \times 5 \times 5$ _____ $= $ _____ $2 \times 3 \times 5^2$ _____

Practice

Write each number in expanded form, then in expanded form using powers of ten.

1. $452 = $ _____ $4 \times 100 + 5 \times 10 + 2 \times 1$ _____ $=$
_____ $4 \times 10^2 + 5 \times 10^1 + 2 \times 10^0$ _____

2. $2,684 = $ _____ $2 \times 1,000 + 6 \times 100 + 8 \times 10 + 4 \times 1$ _____ $=$
_____ $2 \times 10^3 + 6 \times 10^2 + 8 \times 10^1 + 4 \times 10^0$ _____

3. $0.37 = $ _____ $3 \times 0.1 + 7 \times 0.01$ _____
_____ $3 \times 10^{-1} + 7 \times 10^{-2}$ _____

4. $0.683 = $ _____ $6 \times 0.1 + 8 \times 0.01 + 3 \times 0.001$ _____ $=$
_____ $6 \times 10^{-1} + 8 \times 10^{-2} + 3 \times 10^{-3}$ _____

5. $49.35 = $ _____ $4 \times 10 + 9 \times 1 + 3 \times 0.1 + 5 \times 0.01$ _____ $=$
_____ $4 \times 10^1 + 9 \times 10^0 + 3 \times 10^{-1} + 5 \times 10^{-2}$ _____

6. $724.26 = $ _____ $7 \times 100 + 2 \times 10 + 4 \times 1 + 2 \times 0.1 + 6 \times 0.01$ _____ $=$
_____ $7 \times 10^2 + 2 \times 10^1 + 4 \times 10^0 + 2 \times 10^{-1} + 6 \times 10^{-2}$ _____

Write each number in standard form.

7. $7 \times 10^2 + 8 \times 10^1 + 4 \times 10^0 = $ _____ 784 _____

8. $6 \times 10^4 + 9 \times 10^2 + 3 \times 10^1 + 7 \times 10^0 = $ _____ $60,937$ _____

9. $2 \times 10^0 + 5 \times 10^{-1} + 8 \times 10^{-2} + 4 \times 10^{-3} = $ _____ 2.584 _____

10. $8 \times 10^2 + 9 \times 10^0 + 6 \times 10^{-1} + 3 \times 10^{-2} = $ _____ 809.63 _____

Practice

Write each number in expanded form, then in expanded form using powers of ten.

1. $873 = $ _____ $8 \times 100 + 7 \times 10 + 3 \times 1$ _____ $=$
_____ $8 \times 10^2 + 7 \times 10^1 + 3 \times 10^0$ _____

2. $5,104 = $ _____ $5 \times 1,000 + 5 \times 100 + 4 \times 1$ _____ $=$
_____ $5 \times 10^3 + 1 \times 10^2 + 4 \times 10^0$ _____

3. $0.692 = $ _____ $6 \times 0.1 + 9 \times 0.01 + 2 \times 0.001$ _____ $=$
_____ $6 \times 10^{-1} + 9 \times 10^{-2} + 2 \times 10^{-3}$ _____

4. $7.24 = $ _____ $7 \times 1 + 2 \times 0.1 + 4 \times 0.01$ _____ $=$
_____ $7 \times 10^0 + 2 \times 10^{-1} + 4 \times 10^{-2}$ _____

5. $189.56 = $ _____ $1 \times 100 + 8 \times 10 + 9 \times 1 + 5 \times 0.1 + 6 \times 0.01$ _____ $=$
_____ $1 \times 10^2 + 8 \times 10^1 + 9 \times 10^0 + 5 \times 10^{-1} + 6 \times 10^{-2}$ _____

6. $302.48 = $ _____ $3 \times 100 + 2 \times 1 + 4 \times 0.1 + 8 \times 0.01$ _____ $=$
_____ $3 \times 10^2 + 2 \times 10^0 + 4 \times 10^{-1} + 8 \times 10^{-2}$ _____

Write each number in standard form.

7. $7 \times 10^3 + 8 \times 10^2 + 9 \times 10^1 + 4 \times 10^0 = $ _____ $7,894$ _____

8. $5 \times 10^4 + 3 \times 10^3 + 4 \times 10^1 + 5 \times 10^0 = $ _____ $53,045$ _____

9. $4 \times 10^1 + 7 \times 10^0 + 3 \times 10^{-1} + 8 \times 10^{-2} + 1 \times 10^{-3} = $ _____ 47.381 _____

10. $8 \times 10^{-1} + 3 \times 10^{-2} + 4 \times 10^{-3} + 7 \times 10^{-4} = $ _____ 0.8347 _____

Order Fractions: Lesson 1A

Practice

Write the numerator or the denominator to make equivalent fractions.

1. $\frac{2}{3} = \frac{4}{6} = \frac{6}{9} = \frac{8}{12}$ 2. $\frac{5}{8} = \frac{10}{16} = \frac{15}{24} = \frac{20}{32}$

Compare the shaded part of each of these fractions, and write a true statement using $<$, $>$, or $=$.

3. $\frac{1}{5} < \frac{1}{3}$

4. $\frac{2}{4} > \frac{2}{6}$

Rewrite the fractions in each pair so they have the same denominator, and write a true statement using $<$, $>$, or $=$.

5. $\frac{1}{4}$ $\frac{1}{3}$ $\frac{3}{12} < \frac{4}{12}$

6. $\frac{5}{12}$ $\frac{3}{10}$ $\frac{50}{120} > \frac{36}{120}$

Compare the fractions using cross multiplication, and write a true statement using $<$, $>$, or $=$.

7. $\frac{6}{12}$ $\frac{2}{4}$ $6 \times \frac{4}{24} \;?\; 12 \times \frac{2}{24}$ $=$

8. $\frac{3}{5}$ $\frac{3}{7}$ $3 \times \frac{7}{21} \;?\; 5 \times \frac{3}{15}$ $>$

9. $\frac{1}{8}$ $\frac{3}{11}$ $1 \times \frac{11}{11} \;?\; 8 \times \frac{3}{24}$ $<$

10. $\frac{3}{10}$ $\frac{2}{6}$ $3 \times \frac{6}{18} \;?\; 10 \times \frac{2}{20}$ $<$

Order Fractions: Lesson 1B

Practice

Write the numerator or the denominator to make equivalent fractions.

1. $\frac{4}{5} = \frac{8}{10} = \frac{12}{15} = \frac{16}{20}$ 2. $\frac{3}{7} = \frac{6}{14} = \frac{9}{21} = \frac{12}{28}$

Compare the shaded part of each of these fractions, and write a true statement using $<$, $>$, or $=$.

3. $\frac{3}{6} > \frac{3}{8}$

4. $\frac{2}{8} = \frac{1}{4}$

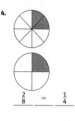

Rewrite the fractions in each pair so they have the same denominator, and write a true statement using $<$, $>$, or $=$.

5. $\frac{2}{5}$ $\frac{2}{3}$ $\frac{6}{15} < \frac{10}{15}$

6. $\frac{5}{8}$ $\frac{2}{5}$ $\frac{25}{40} > \frac{16}{40}$

Compare the fractions using cross multiplication, and write a true statement using $<$, $>$, or $=$.

7. $\frac{2}{9}$ $\frac{2}{7}$ $2 \times \frac{7}{14} \;?\; 9 \times \frac{2}{18}$ $<$

8. $\frac{4}{9}$ $\frac{5}{7}$ $4 \times \frac{9}{36} \;?\; 7 \times \frac{5}{35}$ $>$

Order Fractions: Lesson 2A

Practice

Use mental math or play money to find the answer to each question. Write your answers.

1. How many pennies are in $\frac{7}{10}$ of 100?   **70**

2. $\frac{7}{10} = \frac{70}{100}$

3. What are the decimal equivalents?   **0.7**   **0.70**

Write each as a part of 100 using fraction and decimal form.

4. 32 pennies $\frac{32}{100}$ **0.32**

5. 16 pennies $\frac{16}{100}$ **0.16**

6. 65 pennies $\frac{65}{100}$ **0.65**

7. 83 pennies $\frac{83}{100}$ **0.83**

Write each fraction as a decimal.

8. $\frac{3}{10}$ **0.3**

9. $\frac{38}{100}$ **0.38**

10. $\frac{47}{100}$ **0.47**

11. $\frac{9}{1}$ **9**

12. $\frac{568}{1000}$ **0.568**

13. $\frac{19}{1000}$ **0.019**

14. $\frac{9}{100}$ **0.09**

15. $\frac{4}{5}$ **0.8**

Order Fractions: Lesson 2B

Practice

Use mental math or play money to find the answer to each question. Write your answers.

1. How many pennies are in $\frac{3}{10}$ of 100?   **30**

2. $\frac{3}{10} = \frac{30}{100}$

3. What are the decimal equivalents?   **0.3**   **0.30**

Write each as a part of 100 using fraction and decimal form.

4. 36 pennies $\frac{36}{100}$ **0.36**

5. 52 pennies $\frac{52}{100}$ **0.52**

6. 75 pennies $\frac{75}{100}$ **0.75**

7. 91 pennies $\frac{91}{100}$ **0.91**

Write each fraction as a decimal.

8. $\frac{6}{10}$ **0.6**

9. $\frac{4}{100}$ **0.04**

10. $\frac{82}{100}$ **0.82**

11. $\frac{3}{1000}$ **0.003**

12. $\frac{17}{1000}$ **0.017**

13. $\frac{451}{1000}$ **0.451**

14. $\frac{66}{100}$ **0.66**

15. $\frac{1}{4}$ **0.25**

Algebra Readiness • Practice Answers   215

Name _____ Date _____

Order Fractions: Lesson 3A

Practice

Write the answer to each question. You can draw models to help you find percents.

20% of 35

1. 20% is the same as the fraction $\frac{1}{5}$.
2. The fraction, $\frac{1}{5}$, of 35 is __7__.
3. 20% of 35 is __7__.

25% of 40

4. 25% is the same as the fraction $\frac{1}{4}$.
5. The fraction, $\frac{1}{4}$, of 40 is __10__.
6. 25% of 40 is __10__.

10% of 80

7. 10% is the same as the decimal __0.1__.
8. The decimal, __0.1__, of 80 is __8__.
9. 10% of 80 is __8__.

Complete the chart below so the percent, fraction, and decimal in each row equal each other.

	Percent	Fraction	Decimal
10.	80%	$\frac{80}{100}$	0.8
11.	50%	$\frac{5}{10}$	0.5
12.	30%	$\frac{3}{10}$	0.3
13.	5%	$\frac{5}{100}$	0.05
14.	15%	$\frac{15}{100}$	0.15

Name _____ Date _____

Order Fractions: Lesson 3B

Practice

Write the answer to each question. You can draw models to help you find percents.

10% of 40

1. 10% is the same as the fraction $\frac{1}{10}$.
2. The fraction, $\frac{1}{10}$, of 40 is __4__.
3. 10% of 40 is __4__.

50% of 88

4. 50% is the same as the fraction $\frac{1}{2}$.
5. The fraction, $\frac{1}{2}$, of 88 is __44__.
6. 50% of 88 is __44__.

30% of 60

7. 30% is the same as the decimal __0.3__.
8. The decimal, __0.3__, of 60 is __18__.
9. 30% of 60 is __18__.

Complete the chart below so the percent, fraction, and decimal in each row equal each other.

	Percent	Fraction	Decimal
10.	70%	$\frac{70}{100}$	0.7
11.	40%	$\frac{4}{10}$	0.4
12.	25%	$\frac{25}{100}$	0.25
13.	9%	$\frac{9}{100}$	0.09
14.	62%	$\frac{62}{100}$	0.62

Name _____ Date _____

Order Fractions: Lesson 4A

Practice

Estimate where each number is on the number line. Label each with a point and the given value.

1. 0.3 and 0.75

0   0.3 $\frac{1}{2}$ 0.75   1

2. 20% and 65%

0   20% $\frac{1}{2}$ 65%   1

3. $\frac{2}{5}$ and $\frac{5}{8}$

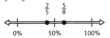

0%   50%   100%

Name a percentage between the given values. Record it on the number line.

4.

12%   45%

any number between 12% and 45%

5.

56%   66%

any number between 56% and 66%

6.

0.4   0.6

any number between 40% and 60%

Name _____ Date _____

Order Fractions: Lesson 4B

Practice

Estimate where each number is on the number line. Label each with a point and the given value.

1. 0.45 and 0.8

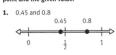

0.45   0.8

0    $\frac{1}{2}$ 1

2. 30% and 85%

0   30% $\frac{1}{2}$ 85% 1

3. $\frac{1}{5}$ and $\frac{3}{4}$

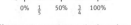

0% $\frac{1}{5}$ 50% $\frac{3}{4}$ 100%

Name a percentage between the given values. Record it on the number line.

4.

35%   60%

any number between 35% and 60%

5.

62%   72%

any number between 62% and 72%

6.

0.44   0.84

any number between 44% and 84%

Adding and Subtracting Rational Numbers: Lesson 1A

Practice

Use the number line to help you solve each problem.

1. $12 - 4 = \underline{8}$

2. $-5 + -8 = \underline{-13}$

3. $5 + -2 = \underline{3}$

4. $-7 + 9 = \underline{2}$

5. $-3 - 6 = \underline{-9}$

6. $3 - -2 = \underline{5}$

7. $8 - 14 = \underline{-6}$

8. $7 - -2 = \underline{9}$

9. $7 + -3 = \underline{4}$

10. $-2 + -8 = \underline{-10}$

11. $-6 - -5 = \underline{-1}$

12. $8 - 2 = \underline{6}$

Adding and Subtracting Rational Numbers: Lesson 1B

Practice

Use the number line to help you solve each problem.

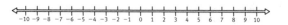

1. $10 - 3 = \underline{7}$

2. $-6 + -2 = \underline{-8}$

3. $4 + -7 = \underline{-3}$

4. $-5 + 10 = \underline{5}$

5. $-5 - 6 = \underline{-11}$

6. $5 + -3 = \underline{2}$

7. $3 - 12 = \underline{-9}$

8. $12 - -4 = \underline{16}$

9. $10 + -3 = \underline{7}$

10. $-3 + -10 = \underline{-13}$

11. $-8 - -2 = \underline{-6}$

12. $-9 + 9 = \underline{0}$

Adding and Subtracting Rational Numbers: Lesson 2A

Practice

Use grids to help you solve the addition and subtraction problems. Write your answers.

1. The basketball team has played $\frac{11}{20}$ of their scheduled games. What part of their scheduled games do they still have left to play?

 $\underline{\frac{9}{20}}$ of the games

2. It rained $\frac{1}{8}$ inch on Monday and $\frac{5}{8}$ inch on Tuesday. What was the total rainfall on these two days?

 $\underline{\frac{6}{8} \text{ or } \frac{3}{4}}$ inch

3. Tia has finished $\frac{3}{4}$ of her math homework problems. What part of her assignment does she still have to complete? Draw your own grid.

 $\underline{\frac{1}{4}}$ of the assignment

4. Ruel has $\frac{3}{4}$ cup of milk in his measuring cup to make a frittata. The recipe says to use $\frac{1}{4}$ cup in the egg mixture and save the rest for the cheese layer. How much milk will be left in the measuring cup after Ruel makes the egg mixture? Draw your own grid.

 $\underline{\frac{2}{4} \text{ or } \frac{1}{2}}$ cup

Adding and Subtracting Rational Numbers: Lesson 2B

Practice

Use grids to help you solve the addition and subtraction problems. Write your answers.

1. Duan, the basketball team manager, is responsible for washing the team uniforms. There are 14 uniforms to be washed this week. He has already washed $\frac{9}{14}$ of the uniforms. What fraction of the uniforms does he still have to wash?

 $\underline{\frac{5}{14}}$ of the uniforms

2. It rained $\frac{7}{8}$ inch on Monday and $\frac{3}{8}$ inch on Tuesday. How much more did it rain on Monday than on Tuesday?

 $\underline{\frac{4}{8} \text{ or } \frac{1}{2}}$ inch

3. The decorating committee finished $\frac{3}{10}$ of the table decorations for the upcoming festival on Tuesday and $\frac{4}{10}$ of the table decorations on Wednesday. What part of the decorations have they finished on these two days? Draw your grid.

 $\underline{\frac{7}{10}}$ of the decorations

4. Ling has $\frac{7}{8}$ quart of motor oil. He needs to use $\frac{5}{8}$ quart to mix with gasoline for his riding lawn mower. What part of a quart of oil will he still have left? Draw your own grid.

 $\underline{\frac{2}{8} \text{ or } \frac{1}{4}}$ quart

Adding and Subtracting Rational Numbers: Lesson 3A

Name _____ Date _____

Practice

Use the grid to add or subtract the decimals.

1. 0.57 + 0.24 = __0.81__

2. 0.88 − 0.26 = __0.62__

3. 0.63 − 0.45 = __0.18__

4. 0.71 − 0.39 = __0.32__

Add or subtract the decimals.

5. 0.36 + 0.08 = __0.44__

6. −0.24 + −0.35 = __−0.59__

7. 0.57 − 0.26 = __0.31__

8. 0.52 + −0.78 = __−0.26__

9. 0.49 + −0.16 = __0.33__

10. −0.51 − −0.98 = __0.47__

Adding and Subtracting Rational Numbers: Lesson 3B

Name _____ Date _____

Practice

Use the grid to add or subtract the decimals.

1. 0.68 + 0.13 = __0.81__

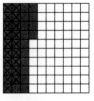

2. 0.74 − 0.27 = __0.47__

3. 0.34 − 0.19 = __0.15__

4. 0.52 − 0.08 = __0.44__

Add or subtract the decimals.

5. 0.63 + 0.28 = __0.91__

6. −0.07 + −0.35 = __−0.42__

7. 0.49 − 0.33 = __0.16__

8. 0.43 + −0.69 = __−0.26__

9. 0.42 + −0.79 = __−0.37__

10. −0.32 − −0.73 = __0.41__

Adding and Subtracting Rational Numbers: Lesson 4A

Name _____ Date _____

Practice

Solve the equations.

1. 3^2 = __9__

2. 8^0 = __1__

3. 16^0 = __1__

4. 10^3 = __1,000__

5. 1^{23} = __1__

6. 5^3 = __125__

7. 27^1 = __27__

8. 5^1 = __5__

9. 1^2 = __1__

10. 7^2 = __49__

11. 4^3 = __64__

12. 11^2 = __121__

Adding and Subtracting Rational Numbers: Lesson 4B

Name _____ Date _____

Practice

Solve the equations.

1. 6^2 = __36__

2. 4^0 = __1__

3. 59^0 = __1__

4. 10^4 = __10,000__

5. 3^3 = __27__

6. 13^2 = __169__

7. 23^1 = __23__

8. 17^1 = __17__

9. 7^2 = __49__

10. 1^4 = __1__

11. 2^3 = __8__

12. 7^0 = __1__

Multiplying and Dividing Rational Numbers: Lesson 1A

Practice

Multiply or divide.

1. $-14 \times -5 =$ __70__
2. $-63 \div -7 =$ __9__
3. $45 \div -9 =$ __-5__
4. $42 \times 16 =$ __672__
5. $28 \times -7 =$ __-196__

6. $-56 \div 8 =$ __-7__
7. $96 \div 12 =$ __8__
8. $200 \div -40 =$ __-5__
9. $-8 \times 34 =$ __-272__
10. $-30 \times -12 =$ __360__

Decide whether to multiply or divide in the following situations. Explain why you chose the operations you chose. Write your answers.

11. The stock for a company dropped 48 points over the last 6 days. On average, how many points did the stock drop each day?

-48 __÷__ $6 =$ __-8__

I chose this operation because __Sample answer: the loss was split over several days__

The stock dropped __8__ points each day.

12. Sixteen students in the drama club have assembled programs for the school play. Each member assembled 25 programs. How many programs have they assembled altogether?

16 __×__ $25 =$ __400__

I chose this operation because __Sample answer: each student was responsible for__ part of the project and you need to put the whole project together

They assembled __400__ programs.

Multiplying and Dividing Rational Numbers: Lesson 1B

Practice

Multiply or divide.

1. $45 \div -9 =$ __-5__
2. $100 \div 4 =$ __25__
3. $-12 \times 8 =$ __-96__
4. $24 \times -2 =$ __-48__
5. $-36 \div -12 =$ __3__

6. $35 \times 12 =$ __420__
7. $-96 \div 3 =$ __-32__
8. $-72 \div -9 =$ __8__
9. $-22 \times -5 =$ __110__
10. $-42 \times -3 =$ __126__

Decide whether to multiply or divide in the following situations. Explain why you chose the operations you chose. Write your answers.

11. Marsa practiced her clarinet an average of 35 minutes each day last week. How many minutes did she practice altogether last week?

35 __×__ $7 =$ __245__

I chose this operation because __Sample answer: you need to put the time together__ for the 7 days last week.

She practiced __245__ minutes last week.

12. The average low temperature dropped 30 degrees over the last 6 days. On average, how many degrees did the low temperature drop each day?

-30 __÷__ $6 =$ __-5__

I chose this operation because __Sample answer: because the decrease was spread__ over several days

The low temperature dropped an average of __5__ degrees each day.

Multiplying and Dividing Rational Numbers: Lesson 2A

Practice

Reduce each fraction by dividing the numerator and the denominator by a common factor. Write your answer.

1. $\frac{12}{20} =$ $\frac{3}{5}$

The common factor that I divided the denominator and numerator by was __4__

2. $\frac{84}{72} =$ $\frac{7}{6}$

The common factor that I divided the denominator and numerator by was __12__

3. $\frac{35}{14} =$ $\frac{5}{2}$

The common factor that I divided the denominator and numerator by was __7__

4. $\frac{15}{42} =$ $\frac{5}{14}$

The common factor that I divided the denominator and numerator by was __3__

Multiply or divide the fractions. Write your answer as a reduced fraction.

5. $\frac{4}{3} \times \frac{9}{16} =$ $\frac{3}{4}$

6. $\frac{2}{5} \div \frac{5}{6} =$ $\frac{12}{25}$

7. $25 \div \frac{1}{5} =$ $\frac{125}{1}$ or 125

8. $\frac{7}{10} \times \frac{15}{21} =$ $\frac{1}{2}$

9. $\frac{5}{3} \div \frac{7}{9} =$ $\frac{15}{7}$ or $2\frac{1}{7}$

10. $35 \times \frac{3}{7} =$ $\frac{15}{1}$ or 15

11. $\frac{6}{7} \times \frac{21}{3} =$ $\frac{6}{1}$ or 6

12. $\frac{10}{3} \div 6 =$ $\frac{5}{9}$

Multiplying and Dividing Rational Numbers: Lesson 2B

Practice

Reduce each fraction by dividing the numerator and the denominator by a common factor. Write your answer.

1. $\frac{18}{30} =$ $\frac{3}{5}$

The common factor that I divided the denominator and numerator by was __6__

2. $\frac{32}{24} =$ $\frac{4}{3}$

The common factor that I divided the denominator and numerator by was __8__

3. $\frac{72}{12} =$ $\frac{6}{1}$ or 6

The common factor that I divided the denominator and numerator by was __12__

4. $\frac{45}{81} =$ $\frac{5}{9}$

The common factor that I divided the denominator and numerator by was __9__

Multiply or divide the fractions. Write your answer as a reduced fraction.

5. $\frac{3}{8} \times \frac{4}{15} =$ $\frac{1}{10}$

6. $\frac{2}{5} \div \frac{5}{8} =$ $\frac{16}{25}$

7. $20 \div \frac{1}{2} =$ $\frac{40}{1}$ or 40

8. $\frac{6}{5} \times \frac{25}{18} =$ $\frac{5}{3}$ or $1\frac{2}{3}$

9. $\frac{3}{4} \div \frac{12}{5} =$ $\frac{5}{16}$

10. $20 \times \frac{3}{8} =$ $\frac{15}{2}$ or $7\frac{1}{2}$

11. $\frac{2}{5} \times 10 =$ $\frac{4}{1}$ or 4

12. $\frac{2}{3} \div 12 =$ $\frac{1}{18}$

Algebra Readiness • Practice Answers 219

Name _____ Date _____

Multiplying and Dividing Rational Numbers: Lesson 3A

Practice

Use the grid to multiply the decimals.

1. $0.4 \times 0.6 =$ __0.24__

2. $0.5 \times 0.5 =$ __0.25__

3. $0.2 \times 0.7 =$ __0.14__

4. $0.3 \times 0.9 =$ __0.27__

5. $0.8 \times 0.1 =$ __0.08__

6. $0.7 \times 0.6 =$ __0.42__

Multiply the decimals. Show your work, and write your answer.

7. Marta can jog 1 mile in 4.5 minutes. At this rate, how long will it take her to jog 6.2 miles? __27.9 minutes__

8. A farmer can harvest 10.5 acres of wheat in 1 hour. How many acres can he harvest in 8.5 hours? __89.25 acres__

Name _____ Date _____

Multiplying and Dividing Rational Numbers: Lesson 3B

Practice

Use the grid to multiply the decimals.

1. $0.5 \times 0.3 =$ __0.15__

2. $0.6 \times 0.1 =$ __0.06__

3. $0.2 \times 0.9 =$ __0.18__

4. $0.3 \times 0.8 =$ __0.24__

5. $0.7 \times 0.7 =$ __0.49__

6. $0.9 \times 0.4 =$ __0.36__

Multiply the decimals. Show your work, and write your answer.

7. An industrial printer can produce 12.5 labels every minute. How many labels can this machine produce in 45 minutes? __562.5 labels__

8. A framing crew can frame a new house in 2.5 days. How many days will it take this crew to frame 7 new houses of the same design? __17.5 days__

Name _____ Date _____

Multiplying and Dividing Rational Numbers: Lesson 4A

Practice

Multiply the decimals by the powers of ten. Write the product.

1. $100 \times 0.658 =$ __65.8__

2. $10 \times 0.237 =$ __2.37__

3. $10 \times 2.467 =$ __24.67__

4. $10,000 \times 14.9 =$ __149,000__

5. $1,000 \times 5.24 =$ __5,240__

6. $100 \times 7.526 =$ __752.6__

7. $0.01 \times 2.888 =$ __0.02888__

8. $0.1 \times 0.247 =$ __0.0247__

Divide the expressions as indicated. Write the result.

9. $0.63 \div 0.7 =$ __0.9__

10. $4.2 \div 0.06 =$ __70__

11. $45.45 \div 0.3 =$ __151.5__

12. $36.36 \div 9 =$ __4.04__

13. $7.69 \div 0.769 =$ __10__

14. $23.4 \div 117 =$ __0.2__

15. $0.144 \div 0.12 =$ __1.2__

16. $1.32 \div 1.2 =$ __1.1__

17. $560 \div 0.2 =$ __2,800__

18. $0.0525 \div 0.25 =$ __0.21__

19. $20 \div 0.8 =$ __25__

20. $0.34 \div 0.2 =$ __1.7__

Chapter 10

Name _____ Date _____

Multiplying and Dividing Rational Numbers: Lesson 4B

Practice

Multiply the decimals by the powers of ten. Write the product.

1. $100 \times 6.728 =$ __672.8__

2. $10 \times 16.52 =$ __165.2__

3. $10 \times 0.089 =$ __0.89__

4. $10,000 \times 9.43 =$ __94,300__

5. $1,000 \times 2.458 =$ __2,458__

6. $100 \times 0.0212 =$ __2.12__

7. $0.1 \times 0.075 =$ __0.0075__

8. $0.01 \times 1.678 =$ __0.01678__

Divide the expressions as indicated. Write the result.

9. $0.56 \div 0.7 =$ __0.8__

10. $8.1 \div 0.03 =$ __270__

11. $16.16 \div 0.4 =$ __40.4__

12. $55.66 \div 1.1 =$ __50.6__

13. $14.7 \div -4.2 =$ __−3.5__

14. $42.23 \div 1.03 =$ __41__

15. $0.324 \div 1.8 =$ __0.18__

16. $1.32 \div 8 =$ __0.165__

17. $425 \div 0.5 =$ __850__

18. $0.0289 \div 0.17 =$ __0.17__

19. $-308 \div -0.77 =$ __400__

20. $129.5 \div -1.75 =$ __−74__

Understanding Rational Numbers: Lesson 1A

Practice

Complete the tables below so the percent, fraction, and decimal in each row equal each other.

	Percent	Fraction
1.	60%	$\frac{3}{5}$
2.	25%	$\frac{1}{4}$
3.	9%	$\frac{9}{100}$
4.	50%	$\frac{1}{2}$

	Percent	Decimal
5.	40%	0.4
6.	18%	0.18
7.	3%	0.03
8.	55%	0.55

	Fraction	Decimal
9.	$\frac{17}{100}$	0.17
10.	$\frac{3}{5}$	0.6
11.	$\frac{9}{10}$	0.9
12.	$\frac{6}{1}$	6.0

Understanding Rational Numbers: Lesson 1B

Practice

Complete the tables below so the percent, fraction, and decimal in each row equal each other.

	Percent	Fraction
1.	35%	$\frac{35}{100}$ or $\frac{7}{20}$
2.	50%	$\frac{1}{2}$
3.	57%	$\frac{57}{100}$
4.	40%	$\frac{2}{5}$

	Percent	Decimal
5.	17%	0.17
6.	2%	0.02
7.	100%	1.0
8.	32%	0.32

	Fraction	Decimal
9.	$\frac{17}{100}$	0.17
10.	$\frac{3}{5}$	0.6
11.	$\frac{5}{100}$	0.05
12.	$\frac{4}{1}$	4.0

Understanding Rational Numbers: Lesson 2A

Practice

Convert the following decimals into fractions. Reduce the fractions to lowest terms. Write your answers.

1. $0.08 = \frac{8}{100}$.
   The reduced form is $\frac{2}{25}$.

2. $0.45 = \frac{45}{100}$.
   The reduced form is $\frac{9}{20}$.

3. $0.42 = \frac{42}{100}$.
   The reduced form is $\frac{21}{50}$.

4. $0.17 = \frac{17}{100}$.
   The reduced form is $\frac{17}{100}$.

5. $0.125 = \frac{125}{1,000}$.
   The reduced form is $\frac{1}{8}$.

6. $1.25 = 1\frac{25}{100}$.
   The reduced form is $1\frac{1}{4}$.

7. $1.4 = 1\frac{4}{10}$.
   The reduced form is $1\frac{2}{5}$.

8. $0.148 = \frac{148}{1,000}$.
   The reduced form is $\frac{37}{250}$.

9. $0.0025 = \frac{25}{10,000}$.
   The reduced form is $\frac{1}{400}$.

10. $0.55 = \frac{55}{100}$.
    The reduced form is $\frac{11}{20}$.

Understanding Rational Numbers: Lesson 2B

Practice

Convert the following decimals into fractions. Reduce the fractions to lowest terms. Write your answers.

1. $0.44 = \frac{44}{100}$.
   The reduced form is $\frac{11}{25}$.

2. $0.35 = \frac{35}{100}$.
   The reduced form is $\frac{7}{20}$.

3. $0.39 = \frac{39}{100}$.
   The reduced form is $\frac{39}{100}$.

4. $0.23 = \frac{23}{100}$.
   The reduced form is $\frac{23}{100}$.

5. $0.175 = \frac{175}{1,000}$.
   The reduced form is $\frac{7}{40}$.

6. $1.75 = 1\frac{75}{100}$.
   The reduced form is $1\frac{3}{4}$.

7. $1.2 = 1\frac{2}{10}$.
   The reduced form is $1\frac{1}{5}$.

8. $0.225 = \frac{225}{1,000}$.
   The reduced form is $\frac{9}{40}$.

9. $0.005 = \frac{5}{1,000}$.
   The reduced form is $\frac{1}{200}$.

10. $1.15 = 1\frac{15}{100}$.
    The reduced form is $1\frac{3}{20}$.

Algebra Readiness • Practice Answers 221

Understanding Rational Numbers: Lesson 3A

Practice

Follow the steps to solve the proportion.

There are 200 students in the middle school who are going on a field trip to the zoo. Fifteen percent of these students have family zoo passes. How many of the students have family zoo passes?

1. What is the unknown in the question?
   __how many students have family zoo passes__

2. What fraction would you set up to represent the percent? __$\frac{15}{100}$__

3. What fraction would you set up to represent the students who have passes (the part) divided by the total number of students going on the trip (the whole)? __$\frac{x}{200}$__

4. What is the proportional equation that you set up? How many students have family zoo passes? $\frac{15}{100} = \frac{x}{200}$
   __30 students have family zoo passes__

Write a proportion for each problem. Write the solution to each problem.

5. There were 40 questions on the last history test. Diego got 85% of these questions right. How many questions did Diego answer correctly?
   $\frac{85}{100} = \frac{x}{40}$ __34 questions__

6. The basketball team scored 90 points in last night's game. Free throws accounted for 18 of those points. What percent of the points in last night's game came from free throws?
   $\frac{x}{100} = \frac{18}{90}$ __20%__

7. The drama club has assembled 150 programs for the upcoming play. This is 40% of all of the programs that need to be put together. How many programs do they need to assemble altogether?
   $\frac{40}{100} = \frac{150}{x}$ __375 programs__

8. Marquis has finished 36 of his math homework problems. There are 40 problems in the assignment. What percent of the problems has he finished?
   $\frac{x}{100} = \frac{36}{40}$ __90%__

Understanding Rational Numbers: Lesson 3B

Practice

Follow the steps to solve the proportion.

There are 160 students in the school who are going on a field trip to the aquarium. Twenty percent of these students have never been to the aquarium before. How many of the students have never been to the aquarium?

1. What is the unknown in the question?
   __how many of the students have never been to the aquarium (the part)__

2. What fraction would you set up to represent the percent? __$\frac{20}{100}$__

3. What fraction would you set up to represent the students who have never been to the aquarium before (the part) divided by the total number of students going on the trip (the whole)? __$\frac{x}{160}$__

4. What is the proportional equation that you set up? How many students have never been to the aquarium before? $\frac{20}{100} = \frac{x}{160}$
   __32 students have never been to the aquarium before.__

Write a proportion for each problem. Write the solution to each problem.

5. There were 30 questions on the last math test. Dora got 90% of these questions right. How many questions did Dora answer correctly?
   $\frac{90}{100} = \frac{x}{30}$ __27 questions__

6. The football team scored 48 points in last Saturday's game. Field goals accounted for 12 of those points. What percent of the points in last Saturday's game came from free throws?
   $\frac{x}{100} = \frac{12}{48}$ __25%__

7. This weekend's long distance bike trip is a 30 mile ride. Ling and Maio plan to ride 18 miles the first day. What percent of the trip do they plan to ride the first day?
   $\frac{x}{100} = \frac{18}{30}$ __60%__

8. Mindy's grade on her math quiz was 85%. She looked at her paper and realized that she had 17 problems correct. How many problems were on the quiz?
   $\frac{85}{100} = \frac{17}{x}$ __20 problems__

Understanding Rational Numbers: Lesson 4A

Practice

Decide whether the percent would make you pay more or less in the following situations, or whether the percent stands alone. Solve the problems. Write *is added to the total*, *is subtracted from the total*, or *stands alone* and the solution.

1. Manuella bought a pair of slacks at the store during a 30% off sale. If the slacks were originally 42 dollars, how much did she pay for them during the sale?
   The percent __is subtracted from the total__. Therefore, the amount she paid during the sale was __$29.40__.

2. Thom and the salesman have agreed on a price for Thom's new car. The price they have agreed upon is 14,500 dollars. In addition, Thom will need to pay the sales tax, title fees, and dealer prep charges which are 12% of the agreed-upon price. How much will Thom pay altogether for the car?
   The percent __is added to the total__. Therefore, the amount Thom will pay for the car is __16,240 dollars__.

3. Ms. Suhara owns shares of stock in a company that has done very well this year. She invested 6,500 dollars in this stock at the beginning of last year. The value of the stock increased 18% during the year. How much money did Ms. Suhara make on the stock last year?
   The percent __stands alone__. Therefore, she earned __1,170 dollars__ on the stock last year.

4. Given the information in Problem 3 above, what was the value of the stock at the beginning of this year?
   The percent __is added to the total__. Therefore, the value of the stock at the beginning of this year was __7,670 dollars__.

Understanding Rational Numbers: Lesson 4B

Practice

Decide whether the percent would make you pay more or less in the following situations, or whether the percent stands alone. Solve the problems. Write *is added to the total*, *is subtracted from the total*, or *stands alone* and the solution.

1. A furniture store offers a 10% discount to its customers who pay cash for their new furniture. The Hollis family is going to buy a new dining room table and chairs priced at 2,450 dollars. How much will they actually pay for the furniture since they have been saving for this purchase and can pay cash?
   The percent __is subtracted from the total__. The price they will pay for the furniture is __2,205 dollars__.

2. Nick and Juanita went out for dinner to celebrate their anniversary. They received very good service and decided to tip 20%. The bill for their dinners originally was 68 dollars. How much will they pay altogether for this dinner?
   The percent __is added to the total__. They will pay __$81.60__ for this dinner.

3. The students at the middle school are having a wrapping-paper sale to raise money for their school's new theater curtain. They will earn 30% of the value of the items that they sell. If the students sell 1,800 dollars during this fund-raising project, how much money will they raise to help to pay for the new curtain?
   The percent __stands alone__. Therefore, they will make __540 dollars__ to help pay for the curtain.

4. The school supply store is having a 15% off sale on all of its top-of-the-line backpacks. How much will a backpack originally priced at 40 dollars sell for during this sale?
   The percent __is subtracted from the total__. Therefore, the price of the backpack during this sale will be __34 dollars__.

Understanding Negative Exponents: Lesson 1A

Practice

Write the following exponential expressions as multiplication problems.
Simplify them by writing them in standard form.

1. 3^2

In expanded form this is ___ 3×3 ___.

Simplified into standard form this is
___ 9 ___

2. 2^5

In expanded form this is
___ $2 \times 2 \times 2 \times 2 \times 2$ ___.

Simplified into standard form this is ___ 32 ___.

3. 7^1

In expanded form this is ___ 7 ___

Simplified into standard form this is
___ 7 ___

4. 6^3

In expanded form this is ___ $6 \times 6 \times 6$ ___

Simplified into standard form this is
___ 216 ___

5. 1^5

In expanded form this is
___ $1 \times 1 \times 1 \times 1 \times 1$ ___

Simplified into standard form this is
___ 1 ___

6. 5^1

In expanded form this is ___ 5 ___.

Simplified into standard form this is
___ 5 ___

7. 10^3

In expanded form this is ___ $10 \times 10 \times 10$ ___

Simplified into standard form this is
___ 1,000 ___

8. 10^5

In expanded form this is
___ $10 \times 10 \times 10 \times 10 \times 10$ ___

Simplified into standard form this is
___ 100,000 ___

9. 12^2

In expanded form this is ___ 12×12 ___

Simplified into standard form this is
___ 144 ___

10. 8^3

In expanded form this is ___ $8 \times 8 \times 8$ ___

Simplified into standard form this is
___ 512 ___

Understanding Negative Exponents: Lesson 1B

Practice

Write the following exponential expressions as multiplication problems.
Simplify them by writing them in standard form.

1. 5^2

In expanded form this is ___ 5×5 ___

Simplified into standard form this is
___ 25 ___

2. 4^4

In expanded form this is
___ $4 \times 4 \times 4 \times 4$ ___

Simplified into standard form this is
___ 256 ___

3. 8^1

In expanded form this is ___ 8 ___

Simplified into standard form this is
___ 8 ___

4. 1^3

In expanded form this is ___ $1 \times 1 \times 1$ ___

Simplified into standard form this is
___ 1 ___

5. 2^4

In expanded form this is ___ $2 \times 2 \times 2 \times 2$ ___

Simplified into standard form this is
___ 16 ___

6. 15^1

In expanded form this is ___ 15 ___.

Simplified into standard form this is
___ 15 ___

7. 10^4

In expanded form this is
___ $10 \times 10 \times 10 \times 10$ ___

Simplified into standard form this is
___ 10,000 ___

8. 10^3

In expanded form this is ___ $10 \times 10 \times 10$ ___

Simplified into standard form this is
___ 1,000 ___

9. 7^3

In expanded form this is ___ $7 \times 7 \times 7$ ___

Simplified into standard form this is
___ 343 ___

10. 12^1

In expanded form this is ___ 12 ___.

Simplified into standard form this is
___ 12 ___

Understanding Negative Exponents: Lesson 2A

Practice

Write the following exponential expressions in expanded form. Simplify
them by writing them in standard form.

1. 3^{-2}

In expanded form this is ___ $\frac{1}{3} \times \frac{1}{3}$ ___.

Simplified into standard form this is
___ $\frac{1}{9}$ ___

2. 4^{-4}

In expanded form this is ___ $\frac{1}{4} \times \frac{1}{4} \times \frac{1}{4} \times \frac{1}{4}$ ___

Simplified into standard form this is
___ $\frac{1}{256}$ ___

3. 8^{-1}

In expanded form this is ___ $\frac{1}{8}$ ___

Simplified into standard form this is
___ $\frac{1}{8}$ ___

4. 1^{-5}

In expanded form this is
___ $\frac{1}{1} \times \frac{1}{1} \times \frac{1}{1} \times \frac{1}{1} \times \frac{1}{1}$ ___

Simplified into standard form this is
___ 1 ___

5. 2^{-3}

In expanded form this is ___ $\frac{1}{2} \times \frac{1}{2} \times \frac{1}{2}$ ___

Simplified into standard form this is
___ $\frac{1}{8}$ ___

6. 12^{-1}

In expanded form this is ___ $\frac{1}{12}$ ___.

Simplified into standard form this is
___ $\frac{1}{12}$ ___

7. 10^{-4}

In expanded form this is
___ $\frac{1}{10} \times \frac{1}{10} \times \frac{1}{10} \times \frac{1}{10}$ ___

Simplified into standard form this is
___ $\frac{1}{10,000}$ ___

8. 10^{-3}

In expanded form this is ___ $\frac{1}{10} \times \frac{1}{10} \times \frac{1}{10}$ ___

Simplified into standard form this is
___ $\frac{1}{1,000}$ ___

9. 5^{-2}

In expanded form this is ___ $\frac{1}{5} \times \frac{1}{5}$ ___.

Simplified into standard form this is
___ $\frac{1}{25}$ ___

10. 9^{-2}

In expanded form this is ___ $\frac{1}{9} \times \frac{1}{9}$ ___

Simplified into standard form this is
___ $\frac{1}{81}$ ___

Understanding Negative Exponents: Lesson 2B

Practice

Write the following exponential expressions in expanded form. Simplify
them by writing them in standard form.

1. 4^{-2}

In expanded form this is ___ $\frac{1}{4} \times \frac{1}{4}$ ___

Simplified into standard form this is
___ $\frac{1}{16}$ ___

2. 3^{-4}

In expanded form this is ___ $\frac{1}{3} \times \frac{1}{3} \times \frac{1}{3} \times \frac{1}{3}$ ___

Simplified into standard form this is
___ $\frac{1}{81}$ ___

3. 9^{-1}

In expanded form this is ___ $\frac{1}{9}$ ___

Simplified into standard form this is
___ $\frac{1}{9}$ ___

4. 1^{-3}

In expanded form this is ___ $\frac{1}{1} \times \frac{1}{1} \times \frac{1}{1}$ ___

Simplified into standard form this is
___ 1 ___

5. 2^{-4}

In expanded form this is
___ $\frac{1}{2} \times \frac{1}{2} \times \frac{1}{2} \times \frac{1}{2}$ ___

Simplified into standard form this is
___ $\frac{1}{16}$ ___

6. 15^{-1}

In expanded form this is ___ $\frac{1}{15}$ ___

Simplified into standard form this is
___ $\frac{1}{15}$ ___

7. 10^{-3}

In expanded form this is ___ $\frac{1}{10} \times \frac{1}{10} \times \frac{1}{10}$ ___

Simplified into standard form this is
___ $\frac{1}{1,000}$ ___

8. 10^{-5}

In expanded form this is
___ $\frac{1}{10} \times \frac{1}{10} \times \frac{1}{10} \times \frac{1}{10} \times \frac{1}{10}$ ___

Simplified into standard form this is
___ $\frac{1}{100,000}$ ___

9. 7^{-3}

In expanded form this is ___ $\frac{1}{7} \times \frac{1}{7} \times \frac{1}{7}$ ___

Simplified into standard form this is
___ $\frac{1}{343}$ ___

10. 4^{-3}

In expanded form this is ___ $\frac{1}{4} \times \frac{1}{4} \times \frac{1}{4}$ ___

Simplified into standard form this is
___ $\frac{1}{64}$ ___

Algebra Readiness • Practice Answers   223

Understanding Negative Exponents: Lesson 3A

Practice

Write the expression modeled by the fraction tiles using a negative exponent. Rewrite the expression as a fraction with a positive exponent.

1. $\frac{1}{8} \times \frac{1}{8} \times \frac{1}{8}$
   $8^{-3} = \frac{1}{8^3}$

2. $\frac{1}{6} \times \frac{1}{6} \times \frac{1}{6} \times \frac{1}{6}$
   $6^{-4} = \frac{1}{6^4}$

3. $\frac{1}{10} \times \frac{1}{10} \times \frac{1}{10} \times \frac{1}{10} \times \frac{1}{10}$
   $10^{-5} = \frac{1}{10^5}$

4. $\frac{1}{4} \times \frac{1}{4} \times \frac{1}{4}$
   $4^{-3} = \frac{1}{4^3}$

5. $\frac{1}{5} \times \frac{1}{5} \times \frac{1}{5} \times \frac{1}{5}$
   $5^{-4} = \frac{1}{5^4}$

6. $\frac{1}{2} \times \frac{1}{2} \times \frac{1}{2} \times \frac{1}{2} \times \frac{1}{2}$
   $2^{-5} = \frac{1}{2^5}$

Understanding Negative Exponents: Lesson 3B

Practice

Write the expression modeled by the fraction tiles using a negative exponent. Rewrite the expression as a fraction with a positive exponent.

1. $\frac{1}{6} \times \frac{1}{6}$
   $6^{-2} = \frac{1}{6^2}$

2. $\frac{1}{10} \times \frac{1}{10} \times \frac{1}{10} \times \frac{1}{10}$
   $10^{-4} = \frac{1}{10^4}$

3. $\frac{1}{3} \times \frac{1}{3} \times \frac{1}{3} \times \frac{1}{3} \times \frac{1}{3}$
   $3^{-5} = \frac{1}{3^5}$

4. $\frac{1}{4} \times \frac{1}{4} \times \frac{1}{4} \times \frac{1}{4} \times \frac{1}{4} \times \frac{1}{4}$
   $4^{-6} = \frac{1}{4^6}$

5. $\frac{1}{2} \times \frac{1}{2} \times \frac{1}{2} \times \frac{1}{2} \times \frac{1}{2} \times \frac{1}{2} \times \frac{1}{2}$
   $2^{-7} = \frac{1}{2^7}$

6. $\frac{1}{12} \times \frac{1}{12} \times \frac{1}{12}$
   $12^{-3} = \frac{1}{12^3}$

Understanding Negative Exponents: Lesson 4A

Practice

Multiply or divide the exponents. Write the result as an exponential expression.

1. $3^5 \times 3^2 = 3^7$

2. $1^{23} \times 1^{15} = 1^{38}$ or 1

3. $5^9 \div 5^3 = 5^6$

4. $5^6 \times 5^7 = 5^{13}$

5. $2^8 \div 2^8 = 2^0$ or 1

6. $2^{-7} \times 2^5 = 2^{-2}$

7. $9^4 \div 9^3 = 9^1$ or 9

8. $4^3 \times 4^7 = 4^{10}$

9. $3^5 \div 3^7 = 3^{-2}$ or $\frac{1}{3^2}$

10. $6^4 \div 6^7 = 6^{-3}$ or $\frac{1}{6^3}$

11. $4^{10} \times 4^{10} = 4^{20}$

12. $6^{-10} \div 6^4 = 6^{-6}$

Understanding Negative Exponents: Lesson 4B

Practice

Multiply or divide the exponents. Write the result as an exponential expression.

1. $4^5 \times 4^8 = 4^{13}$

2. $7^3 \times 7^{15} = 7^{18}$

3. $6^9 \div 6^3 = 6^6$

4. $1^{15} \times 1^{25} = 1^{40}$ or 1

5. $3^7 \div 3^9 = 3^{-2}$ or $\frac{1}{3^2}$

6. $9^{-2} \times 9^{-3} = 9^{-5}$

7. $4^4 \div 4^3 = 4^1$ or 4

8. $84 \times 84 = 8^8$

9. $2^5 \div 2^9 = 2^{-4}$ or $\frac{1}{2^4}$

10. $5^4 \div 5^4 = 5^0$ or 1

11. $6^5 \times 6^5 = 6^{10}$

12. $14^{-2} \div 14^{-2} = 14^0$ or 1

Evaluating Expressions and Writing Equations: Lesson 1A

Practice

Draw parentheses around the operation that should be performed first in order to get the correct answer.

1. $(6 + 12) \div 2 = 9$

2. $30 - (4 \div 2) = 28$

3. $9 - (3 \times 2) = 3$

4. $16 - (4 - 10) = 22$

5. $20 \div (4 + 1) = 4$

6. $(36 \div 2) + 4 = 22$

Write the solution to each expression.

7. $40 - (8 - 12) = \underline{44}$

8. $(12 + 4) \div 4 = \underline{4}$

9. $6 \times (7 + 3) = \underline{60}$

10. $3 + (6 \times 5) = \underline{33}$

11. $(5 \times 5) + 1 = \underline{26}$

12. $6 - (5 + 2) = \underline{-1}$

13. $(2 - 9) + 13 = \underline{6}$

14. $(8 + 2) - 10 = \underline{0}$

15. $(2 - 7) + 5 = \underline{0}$

16. $1 - (8 + 9) = \underline{-16}$

Evaluating Expressions and Writing Equations: Lesson 1B

Practice

Draw parentheses around the operation that should be performed first in order to get the correct answer.

1. $(5 + 8) \times 10 = 130$

2. $8 - ((-4) - 5) = 17$

3. $12 - (5 - 15) = 22$

4. $16 - (4 + 4) = 8$

5. $4 \times (7 + (-2)) = 20$

6. $(12 + 8) \div 4 = 5$

Write the solution to each expression.

7. $(7 \times -3) - 4 = \underline{-25}$

8. $-3 + (-8 \times 2) = \underline{-19}$

9. $11 - (7 + 13) = \underline{-9}$

10. $2 + (8 \times 9) = \underline{74}$

11. $(9 \times 2) + 3 = \underline{21}$

12. $(14 - 4) + 8 = \underline{18}$

13. $5 \times (10 + 5) = \underline{75}$

14. $(5 + 4) \times 3 = \underline{27}$

15. $(9 - 7) + 8 = \underline{10}$

16. $2 - (6 + 8) = \underline{-12}$

Evaluating Expressions and Writing Equations: Lesson 2A

Practice

Use the digits to combine to reach the target number. Write each digit one time in the correct space.

1. Target number: 11
   Digits to combine: 2, 3, 5, 6
   Equation: $\underline{2 + 3 \times 5 - 6}$

2. Target number: 2
   Digits to combine: 2, 3, 4, −5
   Equation: $\underline{(3 \times 4) + (-5 \times 2)}$

3. Target number: 13
   Digits to combine: 2, 3, 4, 7
   Equation: $\underline{(3 - 4) + (7 \times 2)}$

4. Target number: 1
   Digits to combine: 2, 4, 5, 8
   Equation: $\underline{(2 + 8) - (5 + 4)}$

5. Target number: −45
   Digits to combine: −2, −3, 4, 5
   Equation: $\underline{(-2 + -3) \times (4 + 5)}$

6. Target number: −10
   Digits to combine: −2, 3, 5, 6
   Equation: $\underline{(6 \times -2) - (3 - 5)}$

Use the operations to reach the target number. Write each operation one time in the correct space.   Sample answers shown.

7. Target number: 0
   Operations to combine: + + −
   Equation: $(3 \underline{-} 5) \underline{+} (2 \underline{+} 0)$

8. Target number: 12
   Operations to combine: + − ×
   Equation: $(6 \underline{-} 4) \underline{+} (5 \underline{\times} 2)$

9. Target number: 200
   Operations to combine: × × ÷
   Equation: $(20 \underline{\div} 2) \underline{\times} (5 \underline{\times} 4)$

10. Target number: 1
    Operations to combine: + + +
    Equation: $(10 \underline{+} 2) \underline{+} (4 \underline{+} 1)$

Evaluating Expressions and Writing Equations: Lesson 2B

Practice

Use the digits to combine to reach the target number. Write each digit one time in the correct space.

1. Target number: −7
   Digits to combine: 2, 3, 4, 5
   Equation: $\underline{(2 \times 4) - (3 \times 5)}$

2. Target number: −5
   Digits to combine: 1, 2, 4, 8
   Equation: $\underline{(4 - 2) + (1 - 8)}$

3. Target number: 18
   Digits to combine: 3, 0, 12, 2
   Equation: $\underline{(3 \times 2) + (12 + 0)}$

4. Target number: 3
   Digits to combine: 9, 10, 11, 12
   Equation: $\underline{(9 - 12) \times (10 - 11)}$

5. Target number: −11
   Digits to combine: −3, −1, 2, 6
   Equation: $\underline{(-3 - 6) + (-1 \times 2)}$

6. Target number: 8
   Digits to combine: −3, −1, 2, 4
   Equation: $\underline{(-1 + 4) - (-3 - 2)}$

Use the operations to reach the target number. Write each operation one time in the correct space.

7. Target number: −11
   Operations to combine: + − ×
   Equation: $(-8 \underline{\times} 2) \underline{+} (9 \underline{-} 4)$

8. Target number: −4
   Operations to combine: + − ÷
   Equation: $(6 \underline{+} 2) \underline{-} (3 \underline{-} 5)$

9. Target number: −2
   Operations to combine: + − −
   Equation: $(-3 \underline{-} 2) \underline{+} (6 \underline{-} 3)$

10. Target number: −100
    Operations to combine: × × ÷
    Equation: $(-20 \underline{\div} 4) \underline{\times} (10 \underline{\times} 2)$

Algebra Readiness • Practice Answers 225

Evaluating Expressions and Writing Equations: Lesson 3A

Practice

Evaluate the following expressions. Write the solution.

1. $4^2 \times (2 \times 10) + 7 =$ __327__

4. $5^2 + 9 \times (8 - 8) =$ __25__

2. $11 \times (5 - 3) + 4 =$ __26__

5. $0 \times (7 + 11) + 4^2 =$ __16__

3. $13 - (3^2 - 2) + 12 =$ __18__

6. $6 + (3 - 9) \times 2^3 =$ __−42__

Draw a line from the equation to the correct answer.

7. $(4 + 2) - (3 \div 1) =$

8. $(4 \div 2) \times (3 - 1) =$

9. $4 - (2 - 3) \times 1 =$

10. $(4 \times 2) \div (3 + 1) =$

A. 2
B. 3
C. 5
D. 4

Evaluating Expressions and Writing Equations: Lesson 3B

Practice

Evaluate the following expressions. Write the solution.

1. $-5 + -3 \times (-4 - 2) =$ __13__

4. $5^2 \times 2 - (10 \times 5) =$ __0__

2. $(14 - 9) \times 3 - 3 =$ __12__

5. $1^5 \times (4 \times -3) \div 2^2 =$ __−3__

3. $6^2 - (4 \times 3) + 8 =$ __32__

6. $12 - (3 \times 7) + 1^2 =$ __−8__

Draw a line from the equation to the correct answer.

7. $(5 - 4) - (2 - 6) =$

8. $(5 \times 4) \div 2 - 6 =$

9. $5 + (4 \div 2) - 6 =$

10. $(5 + 4) \times 2 \times 6 =$

A. 1
B. 108
C. 5
D. 4

Evaluating Expressions and Writing Equations: Lesson 4A

Practice

Write the expressions and solutions that are described.

1. Luis is 4 years less than twice as old as his sister. If Louis is 12, then how old is his sister?

   $2 = 2s - 4$, where s is his sister's age

   His sister is __8__ years old.

2. If you add 6 to an unknown number and multiply that sum by 7, you get 14. What is the number?

   $14 = 7(n + 6)$, where n is the unknown number

   The unknown number is __−4__.

3. If an unknown number is divided by 8 and then 7 is added to the quotient, the result is 10. What is the number?

   $10 = n \div 8 + 7$, where n is the unknown number

   The unknown number is __24__.

Write the solutions to the expressions that are described. It might be useful to use counters to model the expressions that are described.

4. The cost of a DVD is $4 more than three times the cost of a movie ticket. If the cost of the DVD is $17.50, then what is the price of the movie?

   __$4.50__

5. Marcella earned $16 less in tips than twice what Arvana earned. If Marcella earned $42, then how much did Arvana make in tips?

   __$29__

Evaluating Expressions and Writing Equations: Lesson 4B

Practice

Write the expressions and solutions that are described.

1. If you add 17 to an unknown number, you get one-third of 42. What is the number?

   $42 \div 3 = n + 17$, where n is the unknown number

   The number is __−3__.

2. Angela's age is 3 years less than 4 times her sister's age. If Angela is 13, how old is her sister?

   $13 = 4s - 3$, where s is her sister's age

   Her sister is __4__ years old.

3. If an unknown number is decreased by 10 then that difference is equal to −4. What is the number?

   $n - 10 = -4$, where n is the unknown number

   The unknown number is __6__.

Write the solutions to the expressions that are described. It might be useful to use counters to model the expressions that are described.

4. The cost of a football is $6 less than twice the cost of a soccerball. If the cost of the football is $18, how much does the soccerball cost?

   __$12__

5. If you multiply a number by 7 and then subtract −8, the result is −20? What is the number? __−4__

Using Variables: Lesson 1A

Practice

Draw a line from the expression to the correct input/output.

1. $-b + 3$ **A.** input: 2; output: -3

2. $5c + (-2)$ **B.** input: 0; output: 3

3. $-2m + 1$ **C.** input: 1; output: -3

4. $2p - 5$ **D.** input: 1; output: 3

Write the expression that would give you the outputs desired.

5.

Input	Output
0	0
1	-2
2	-4
3	-6

Expression: $-2x$

7.

Input	Output
0	2
1	3
2	4
3	5

Expression: $x + 2$

6.

Input	Output
0	0
1	1
2	8
3	27

Expression: x^3

8.

Input	Output
0	-5
1	-4
2	-3
3	-2

Expression: $x - 5$

Using Variables: Lesson 1B

Practice

Draw a line from the expression to the correct input/output.

1. $-a + 4$ **A.** input: 2; output: 3

2. $5c - 8$ **B.** input: 0; output: -3

3. $-2n - 3$ **C.** input: 1; output: 3

4. $4p - 5$ **D.** input: 1; output: -3

Write the expression that would give you the outputs desired.

5.

Input	Output
0	4
1	5
2	6
3	7

Expression: $x + 4$

7.

Input	Output
0	0
1	-3
2	-6
3	-9

Expression: $-3x$

6.

Input	Output
0	-2
1	-1
2	0
3	1

Expression: $x - 2$

8.

Input	Output
0	0
1	1
2	4
3	9

Expression: x^2

Using Variables: Lesson 2A

Practice

Write the solution for the variable. It might be useful to use counters to model the situations.

1. $30 \div k = 6$
   $k = 5$

4. $m - 9 = -2$
   $m = 7$

2. $-3e = 15$
   $e = -5$

5. $c - 4 = 10$
   $c = 14$

3. $7 + j = -4$
   $j = -11$

6. $7 + b = -3$
   $b = -10$

Complete the table by writing the missing input or output for the expression $x - 6$.

	Input	Output
7.	-3	-9
8.	0	-6
9.	4	-2
10.	8	2

Using Variables: Lesson 2B

Practice

Write the solution for the variable. It might be useful to use counters to model the situations.

1. $d - 7 = 12$
   $d = 19$

4. $c \div 8 = 4$
   $c = 32$

2. $9 + k = -1$
   $k = -10$

5. $12 - e = 7$
   $e = 5$

3. $-5m = 30$
   $m = -6$

6. $-4y = 0$
   $y = 0$

Complete the table by writing the missing input or output for the expression $x + 5$.

	Input	Output
7.	-7	-2
8.	-2	3
9.	0	5
10.	4	9

Worksheet 1 (top-left)

Practice

Draw a number line to represent the quantities described by the expressions. Sample answers shown

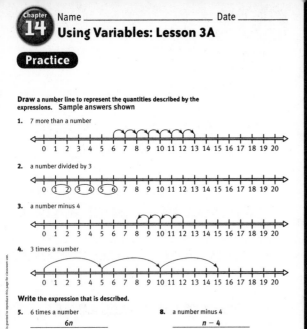

1. 7 more than a number
2. a number divided by 3
3. a number minus 4
4. 3 times a number

Write the expression that is described.

5. 6 times a number
   $6n$

6. 5 more than a number
   $5 + n$ or $n + 5$

7. 8 divided by a number
   $8 \div n$

8. a number minus 4
   $n - 4$

9. 10 less than a number
   $n - 10$

10. a number divided by 7
    $n \div 7$

Worksheet 2 (top-right)

Practice

Draw a number line to represent the quantities described by the expressions. Sample answers shown

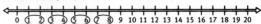

1. 4 times a number
2. a number minus 8
3. a number divided by 4
4. 6 more than a number

Write the expression that is described.

5. 6 more than a number
   $n + 6$ or $6 + n$

6. 12 divided by a number
   $12 \div n$

7. 8 less than a number
   $n - 8$

8. 4 times a number
   $4n$

9. a number divided by −3
   $n \div -3$

10. 11 minus a number
    $11 - n$

Worksheet 3 (bottom-left)

Practice

Draw a line from the description to the correct equation.

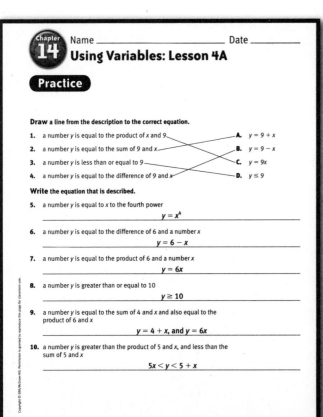

1. a number y is equal to the product of x and 9.
2. a number y is equal to the sum of 9 and x.
3. a number y is less than or equal to 9.
4. a number y is equal to the difference of 9 and x.

A. $y = 9 + x$
B. $y = 9 - x$
C. $y = 9x$
D. $y \le 9$

Write the equation that is described.

5. a number y is equal to x to the fourth power
   $y = x^4$

6. a number y is equal to the difference of 6 and a number x
   $y = 6 - x$

7. a number y is equal to the product of 6 and a number x
   $y = 6x$

8. a number y is greater than or equal to 10
   $y \ge 10$

9. a number y is equal to the sum of 4 and x and also equal to the product of 6 and x
   $y = 4 + x$, and $y = 6x$

10. a number y is greater than the product of 5 and x, and less than the sum of 5 and x
    $5x < y < 5 + x$

Worksheet 4 (bottom-right)

Practice

Draw a line from the description to the correct equation.

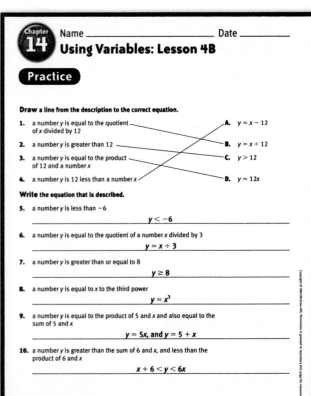

1. a number y is equal to the quotient of x divided by 12
2. a number y is greater than 12
3. a number y is equal to the product of 12 and a number x
4. a number y is 12 less than a number x

A. $y = x - 12$
B. $y = x \div 12$
C. $y > 12$
D. $y = 12x$

Write the equation that is described.

5. a number y is less than −6
   $y < -6$

6. a number y is equal to the quotient of a number x divided by 3
   $y = x \div 3$

7. a number y is greater than or equal to 8
   $y \ge 8$

8. a number y is equal to x to the third power
   $y = x^3$

9. a number y is equal to the product of 5 and x and also equal to the sum of 5 and x
   $y = 5x$, and $y = 5 + x$

10. a number y is greater than the sum of 6 and x, and less than the product of 6 and x
    $x + 6 < y < 6x$

Name _____ Date _____

Two-Variable Equations and Manipulating Symbols: Lesson 1A

Practice

1. Write the *x*-value in the table for the equation $y = 3x - 1$

x	y
−3	−10
−2	−7
−1	−4
0	−1
1	2
2	5
3	8

2. Write the *y*-value in the table for the equation $y = 4 + 3x$

x	y
−4	−8
−3	−5
−2	−2
−1	1
0	4
1	7
2	10

Create a table to represent some solutions for the following equations. Answers will vary.

3. $y = x - 3$

x	y

5. $y = x + 2$

x	y

4. $x = 4$

x	y

6. $y = 2x - 1$

x	y

Name _____ Date _____

Two-Variable Equations and Manipulating Symbols: Lesson 1B

Practice

1. Write the *x*-value in the table for the equation $y = 4 - 2x$

x	y
−3	10
−2	8
−1	6
0	4
1	2
2	0
3	−2

2. Write the *y*-value in the table for the equation $y = 5x + 1$

x	y
−4	−19
−3	−14
−2	−9
−1	−4
0	1
1	6
2	11

Create a table to represent some solutions for the following equations. Answers will vary.

3. $y = x + 3$

x	y

5. $y = -3x - 1$

x	y

4. $y = -1$

x	y

6. $y = -x + 3$

x	y

Name _____ Date _____

Two-Variable Equations and Manipulating Symbols: Lesson 2A

Practice

Draw a line from the equations to the number that makes the statement true.

1. $a = b; a - 3 = b - $ __E__   A. 5

2. $x = y; 1 + x = y + $ __C__   B. −3

3. $j = k; j + 5 = k + $ __A__   C. 1

4. $m = n; -3 - m = $ __B__ $ - n$ D. −5

5. $c = d; 0 + c = d + $ __D__   E. 3

6. $u = v; u + -5 = $ __F__ $ + v$ F. 0

Write the number that makes the statement true.

7. $p = q; p + 2 = q + $ __2__

8. $a = b; a - 4 = b + $ __−4__

9. $r = s; r - 10 = s - $ __10__

10. $m = n; m + 3 = n + (2 + $ __1__ $)$

11. $j = k; -3 - j = $ __−3__ $ - k$

12. $c = d; 7 + c = d + $ __7__

Name _____ Date _____

Two-Variable Equations and Manipulating Symbols: Lesson 2B

Practice

Draw a line from the equations to the number that makes the statement true.

1. $y = z; y + 4 = z + $ __D__   A. −5

2. $d = e; -5 + d = $ __A__ $ + e$ B. 2

3. $p = q; p + 0 = $ __F__ $ + q$ C. 6

4. $s = t; -s - 2 = -t - $ __B__   D. 4

5. $a = c; 6 + a = c + $ __C__   E. 3

6. $r = m; r - 3 = m - $ __E__   F. 0

Write the number that makes the statement true.

7. $a = c; a - 6 = c - $ __6__

8. $r = t; r + 5 = t + (1 + $ __4__ $)$

9. $d = g; d + 12 = $ __12__ $ + g$

10. $b = j; -7 + b = j + $ __−7__

11. $h = k; h - 0 = k - $ __0__

12. $m = w; m + (-2) = w - $ __2__

Two-Variable Equations and Manipulating Symbols: Lesson 3A

Practice

Draw a line from each equation to the number that makes the statement true.

1. $y = z$; $-2 \times y = z \times$ ___D___   **A.** 4

2. $d = e$; $4 \times d =$ ___A___ $\times e$ **B.** 2

3. $p = q$; $-3 \times p =$ ___F___ $\times q$ **C.** −1

4. $s = t$; $s \times 2 = t \times$ ___B___   **D.** −2

5. $a = c$; $-1 \times a = c \times$ ___C___   **E.** −4

6. $r = m$; $-4 \times r = m \times$ ___E___   **F.** −3

Write the number that makes the statement true.

7. $a = c$; $5a =$ ___5___ c 10. $b = j$; $0.3b = j \times$ ___0.3___

8. $r = t$; $-7r =$ ___−7___ t 11. $h = k$; $h \div 4 =$ ___$\frac{1}{4}$___ k

9. $d = g$; $\frac{1}{5}d =$ ___$\frac{1}{5}$___ g 12. $m = w$; $15m = ($ ___5___ $\times 3)w$

Two-Variable Equations and Manipulating Symbols: Lesson 3B

Practice

Draw a line from each equation to the number that makes the statement true.

1. $a = b$; $3 \times a = b \times$ ___D___   **A.** 7

2. $x = y$; $x \times 7 = y \times$ ___A___   **B.** −2

3. $j = k$; $-3j = k \times$ ___F___   **C.** 6

4. $m = n$; $-2 \times m =$ ___B___ $\times n$ **D.** 3

5. $c = d$; $6 \times c = d \times$ ___C___   **E.** 9

6. $u = v$; $u \times 9 =$ ___E___ $\times v$ **F.** −3

Write the number that makes the statement true.

7. $p = q$; $4p =$ ___4___ q 10. $m = n$; $0.4m = m \times$ ___0.4___

8. $a = b$; $-2a =$ ___−2___ b 11. $j = k$; $j \div 2 =$ ___$\frac{1}{2}$___ k

9. $r = s$; $\frac{1}{3}r =$ ___$\frac{1}{3}$___ s 12. $c = d$; $12c = ($ ___4___ $\times 3)d$

Two-Variable Equations and Manipulating Symbols: Lesson 4A

Practice

Write the operation that you selected to help you isolate the variable, and then perform it to both sides. Write the resulting equation.

1. $x - 5 = 10$

   The operation I performed is
   ___adding 5___

   The resulting equation is ___$x = 15$___

2. $-3x = 12$

   The operation I performed is
   ___dividing by −3___

   The resulting equation is ___$x = -4$___

3. $x \div 4 = 20$

   The operation I performed is
   ___multiplying by 4___

   The resulting equation is ___$x = 80$___

4. $12 + x = 15$

   The operation I performed is
   ___subtracting 12___

   The resulting equation is ___$x = 3$___

Write the operations that you selected to help you isolate the variable, and then perform them to both sides. Write the resulting equations.

5. $4x - 7 = 13$

   The first operation I performed is
   ___adding 7___

   The resulting equation is ___$4x = 20$___

   The next operation I performed is
   ___dividing by 4___

   The resulting equation is ___$x = 5$___

6. $-3x + 6 = -6$

   The first operation I performed is
   ___subtracting 6___

   The resulting equation is ___$-3x = -12$___

   The next operation I performed is
   ___dividing by −3___

   The resulting equation is ___$x = 4$___

Isolate the variable. Write the solution.

7. $x - 8 = 62$ ___$x = 70$___

8. $x \div 4 = 12$ ___$x = 48$___

9. $2x - 11 = 43$ ___$x = 27$___

10. $-5x + 2 = 27$ ___$x = -5$___

Two-Variable Equations and Manipulating Symbols: Lesson 4B

Practice

Write the operation that you selected to help you isolate the variable, and then perform it to both sides. Write the resulting equation.

1. $x + 4 = 11$

   The operation I performed is
   ___subtracting 4___

   The resulting equation is ___$x = 7$___

2. $x \div 3 = -12$

   The operation I performed is
   ___multiplying by 3___

   The resulting equation is ___$x = -36$___

3. $-6x = -24$

   The operation I performed is
   ___dividing by −6___

   The resulting equation is ___$x = 4$___

4. $-9 + x = 2$

   The operation I performed is
   ___adding 9___

   The resulting equation is ___$x = 11$___

Write the operations that you selected to help you isolate the variable, and then perform them to both sides. Write the resulting equations.

5. $-2x + 11 = 17$

   The first operation I performed is
   ___subtracting 11___

   The resulting equation is ___$-2x = 6$___

   The next operation I performed is
   ___dividing by −2___

   The resulting equation is ___$x = -3$___

6. $6x - 4 = 38$

   The first operation I performed is
   ___adding 4___

   The resulting equation is ___$6x = 42$___

   The next operation I performed is
   ___dividing by 6___

   The resulting equation is ___$x = 7$___

Isolate the variable. Write the solution.

7. $x + 10 = 34$ ___$x = 24$___

8. $x \times 7 = -28$ ___$x = -4$___

9. $3x - 1 = 23$ ___$x = 8$___

10. $-2x + 6 = 0$ ___$x = 3$___

230   **Algebra Readiness** • Practice Answers

Practice

Apply the properties to simplify the expressions. Write the resulting expressions.

1. $(19 + 4) + 6$; associative property of addition
$$19 + (4 + 6)$$

2. $12 + (-12)$; inverse property of addition
$$0$$

3. $12 + 87 + 18$; commutative property of addition
$$12 + 18 + 87$$

4. 17×1; identity property of multiplication
$$17$$

5. $\frac{1}{5} \times 5$; inverse property of multiplication
$$1$$

6. $10 \times 12 \times 3$; commutative property of multiplication
$$10 \times 3 \times 12$$

7. $12 \times (2 + 3)$; distributive property
$$12 \times 2 + 12 \times 3$$

8. $(19 \times 5) \times 4$; associative property of multiplication
$$19 \times (5 \times 4)$$

Answer the questions about the way the following problem was solved.

(1) $7(x + 4) = 3x$

(2) $7x + 28 = 3x$

(3) $-7x + 7x + 28 = -7x + 3x$

(4) $28 = -4x$

(5) $-\frac{1}{4} \times 28 = -\frac{1}{4} \times -4x$

(6) $-7 = x$

9. On the right side of the equation, what property did the person use to go from step (5) to step (6)? **inverse property of multiplication**

10. Is $x = 7$ the correct solution? Prove your answer.
yes; $7(-7 + 4) = 7(-3) = -21$ and $3(-7) = -21$

Practice

Apply the properties to simplify the expressions. Write the resulting expressions.

1. $8 \times (5 \times 13)$; associative property of multiplication
$$(8 \times 5) \times 13$$

2. $6 \times 19 \times 5$; commutative property of multiplication
$$6 \times 5 \times 19$$

3. $-29 + 29$; inverse property of addition
$$0$$

4. 143×1; identity property of multiplication
$$143$$

5. $20 \times (5 + 1)$; distributive property
$$20 \times 5 + 20 \times 1$$

6. $15 + 27 + 25$; commutative property of addition
$$15 + 25 + 27$$

7. $8 \times \frac{1}{8}$; inverse property of multiplication
$$1$$

8. $12 + (28 + 34)$; associative property of addition
$$(12 + 28) + 34$$

Answer the questions about the way the following problem was solved.

(1) $-3(x - 8) = x$

(2) $-3x + 24 = x$

(3) $3x - 3x + 24 = 3x + x$

(4) $24 = 4x$

(5) $\frac{1}{4} \times 24 = \frac{1}{4} \times 4x$

(6) $6 = x$

9. On the right side of the equation, what property did the person use to go from step (5) to step (6)? **inverse property of multiplication**

10. Is $x = 6$ the correct solution? Prove your answer.
yes; $-3(6 - 8) = -3(-2) = 6$

Practice

Write the solutions to the two-step equations. Verify your solutions by substituting for x. Show your work.

1. $5x - 2 = 18$ $x = 4$ 　　　$5 \times \underline{4} - 2 = 18$

2. $x \div 4 - 3 = 3$ $x = 24$ 　　　$\frac{24}{4} - 3 = 3$

3. $-3x + 9 = 15$ $x = -2$ 　　　$-3 \times \underline{-2} + 9 = 15$

4. $12 = 2x - 4$ $x = 8$ 　　　$12 = 2 \times \underline{8} - 4$

5. $3 = 7x + 3$ $x = 0$ 　　　$3 = 7 \times \underline{0} + 3$

6. $\frac{1}{2}x - 5 = 4$ $x = 18$ 　　　$\frac{1}{2} \times \underline{18} - 5 = 4$

Practice

Write the solutions to the two-step equations. Verify your solutions by substituting for x. Show your work.

1. $3x + 5 = 17$ $x = 4$ 　　　$3 \times \underline{4} + 5 = 17$

2. $-4x - 3 = 21$ $x = -6$ 　　　$-4 \times \underline{-6} - 3 = 21$

3. $\frac{1}{3}x + 7 = 10$ $x = 9$ 　　　$\frac{1}{3} \times \underline{9} + 7 = 10$

4. $x \div 6 + 4 = 5$ $x = 6$ 　　　$\frac{6}{6} + 4 = 5$

5. $10 = 6 + 4x$ $x = 1$ 　　　$10 = 6 + 4 \times \underline{1}$

6. $6 = 2x - 8$ $x = 7$ 　　　$6 = 2 \times \underline{7} - 8$

Chapter 16

Name _____ **Date** _____

Simplifying and Solving Equations: Lesson 3A

Practice

Write the solutions to the two-step inequalities. Verify your solutions by substituting for *x*. Show your work.

1. $2x - 4 > 10$ **x > 7**   2 _any number > 7_ – 4 > 10

2. $3x + 5 < 17$ **x < 4**   3 _any number < 4_ + 5 < 17

3. $21 \le 6x + 9$ **x ≥ 2**   21 ≤ 6 _any number ≥ 2_ + 9

4. $-10 > 2x + 6$ **x < −8**   −10 > 2 _any number < −8_ + 6

5. $11x + 1 \ge 12$ **x ≥ 1**   11 _any number ≥1_ + 1 ≥ 12

6. $51 < 4x + 7$ **x > 11**   51 < 4 _any number > 11_ + 7

Chapter 16

Name _____ **Date** _____

Simplifying and Solving Equations: Lesson 3B

Practice

Write the solutions to the two-step inequalities. Verify your solutions by substituting for *x*. Show your work.

1. $3x - 8 < 13$ **x < 7**   3 _any number < 7_ – 8 < 13

2. $64 < 5x + 9$ **x > 7**   64 < 5 _any number > 11_ + 9

3. $7x + 3 > 31$ **x > 4**   7 _any number > 4_ + 3 > 31

4. $11 \le 4x - 9$ **x ≥ 5**   11 ≤ 4 _any number ≥ 5_ – 9

5. $28 > 6x - 8$ **x < 6**   28 > 6 _any number < 6_ – 8

6. $9x + 2 \ge 20$ **x ≥ 2**   9 _any number ≥ 2_ + 2 ≥ 20

Chapter 16

Name _____ **Date** _____

Simplifying and Solving Equations: Lesson 4A

Practice

Answer the following questions.

The Crumb and Coffee Shoppe sells 6 of its special muffins for $3.90. At this rate (price), how much would you pay for 10 muffins?

1. What is being asked? _____ **how much would 10 muffins cost?**

2. What is the rate that you need? _____ **cost per muffin**

3. How do you find this rate? _____ **$3.90 ÷ 6**

4. What is this rate? _____ **$0.65 per muffin**

5. How do you use this rate to find out how much 10 muffins cost? _____ **multiply rate by 10**

6. How much would 10 muffins cost? _____ **$6.50**

A trucker drives 248 miles in 4 hours. At this rate how long would it take her to drive 682 miles?

7. What is her rate? _____ **62 miles per hour**

8. How did you find her rate? _____ **divided the miles by the hours**

9. How do you use this rate to find out how long it will take her to drive 682 miles? _____ **divide 682 miles by the rate**

10. At this rate, how long will it take her to drive 682 miles? _____ **11 hours**

Chapter 16

Name _____ **Date** _____

Simplifying and Solving Equations: Lesson 4B

Practice

Answer the following questions.

Maria, Manuella's office assistant, can type 5 pages of financial reports in 2 hours. At this rate, how many pages of financial reports can she type in 5 hours?

1. What is being asked? _____ **how much can she type in 5 hours?**

2. What is the rate that you need? _____ **page per hour**

3. How do you find this rate? _____ **5 ÷ 2**

4. What is this rate? _____ $2\frac{1}{2}$ **or 2.5 pages per hour**

5. How do you use this rate to find out how many pages she can type in 5 hours? _____ **multiply rate by 5**

6. How many pages can she type in 5 hours? _____ $12\frac{1}{2}$ **or 12.5 pages**

Chef Andre can make 15 of his special gourmet salads in 12 minutes. At this rate, how long would it take him to prepare 40 of these salads?

7. What is his rate? _____ $1\frac{1}{4}$ **salad per minute**

8. How did you find his rate? _____ **divided salads by minutes**

9. How do you use this rate to find out how long it will take him to make 40 salads? _____ **multiply the rate by 40**

10. At this rate, how long will it take him to make 40 salads? _____ **50 minutes**

Chapter 17

Name _____ Date _____

Points in the Coordinate Plane: Lesson 1A

Practice

Plot each of the points on the grid provided. Identify the points by their coordinates.

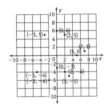

1. (2, 5)
2. (2, −4)
3. (0, −3)
4. (0, 6)
5. (−2, −6)
6. (7, 1)
7. (3, −5)
8. (5, 0)
9. (−3, 5)
10. (−5, −4)

11. Describe how to get from Point 1 above to Point 5. Be sure to describe left-right movement first and then up-down movement.

   move to the left 4 spaces and then down 11 spaces

12. Describe how to get from Point 10 above to Point 4.

   move 5 spaces to the right and then up 10 spaces

13. Describe how to get from Point 3 above to Point 4.

   stay on the y-axis and move up 9 spaces

14. Describe how to get from Point 7 above to Point 8. Be sure to describe left-right movement first and then up-down movement.

   move to the right 2 spaces and then up 5 spaces

Name _____ Date _____

Points in the Coordinate Plane: Lesson 1B

Practice

Plot each of the points on the grid provided. Identify the points by their coordinates.

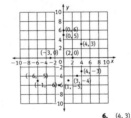

1. (0, 5)
2. (1, −5)
3. (4, −3)
4. (0, 6)
5. (−1, −6)
6. (4, 3)
7. (3, −4)
8. (2, 0)
9. (−3, 0)
10. (−6, −5)

11. Describe how to get from Point 4 above to Point 8. Be sure to describe left-right movement first and then up-down movement.

   move to the right 2 spaces and then down 6 spaces

12. Describe how to get from Point 1 above to Point 5.

   move 1 space to the left and then down 11 spaces

13. Describe how to get from Point 3 above to Point 6.

   stay on the x = 4 line and move up 6 spaces

14. Describe how to get from Point 8 above to Point 10.

   move to the left 8 spaces and then down 5 spaces

Name _____ Date _____

Points in the Coordinate Plane: Lesson 2A

Practice

Graph these points on a grid: A (−3, −1); B (−2, 2); C (3, 2); and D (7, −1). Connect the dots to make a geometric shape. Identify the shape you have made and answer the questions.

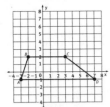

1. The shape is a ___ trapezoid ___
2. In which quadrant is point A located? ___ Quadrant III ___
3. In which quadrant is point B located? ___ Quadrant II ___
4. In which quadrant is point C located? ___ Quadrant I ___
5. In which quadrant is point D located? ___ Quadrant IV ___
6. What is the relationship between side BC and side AD? ___ they are parallel ___

Write the quadrant in which each point is located. You can use a coordinate grid to help you identify the location. If the point is not in a quadrant, list the axis or origin on which it lies.

7. (2, 6) ___ Quadrant I ___
8. (−5, 4) ___ Quadrant II ___
9. (1, −10) ___ Quadrant IV ___
10. (−6, −9) ___ Quadrant III ___
11. (5, 0) ___ on the x-axis ___
12. (0, 0) ___ origin ___
13. (−4, −3) ___ Quadrant III ___
14. (0, −2) ___ on the y-axis ___

Name _____ Date _____

Points in the Coordinate Plane: Lesson 2B

Practice

Graph these points on a grid: A (3, −1); B (1, 3); C (−4, 3); D (−5, 0) and E (0, −3). Connect the dots to make a geometric shape. Identify the shape you have made and answer the questions.

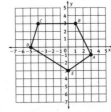

1. The shape is a ___ pentagon ___
2. Describe the location of point A. ___ Quadrant IV ___
3. Describe the location of point B. ___ Quadrant I ___
4. Describe the location of point C. ___ Quadrant II ___
5. Describe the location of point D. ___ on the x-axis ___
6. Describe the location of point E. ___ on the y-axis ___

Write the quadrant in which each point is located. You can use a coordinate grid to help you identify the location. If the point is not in a quadrant, list the axis or origin on which it lies.

7. (−2, 5) ___ Quadrant II ___
8. (0, 0) ___ origin ___
9. (−1, −8) ___ Quadrant III ___
10. (6, −3) ___ Quadrant IV ___
11. (−4, 0) ___ on the x-axis ___
12. (−2, 2) ___ Quadrant II ___
13. (2, 5) ___ Quadrant I ___
14. (0, −7) ___ on the y-axis ___

Name _____ **Date** _____

Points in the Coordinate Plane:
Lesson 3A

Practice

Plot the given points on the coordinate grid. Use a ruler and draw a line that goes through the origin and the point you have placed on the grid.

1. (3, 1)

3. (−4, −1)

2. (4, 0)

4. (2, −1)

Plot the given points on the coordinate grid. Use a ruler and draw a line that goes through the two points you have placed on the grid.

5. (1, −3) and (−2, 1)

7. (−3, 4) and (0, 3)

6. (0, 4) and (0, 1)

8. (−2, −4) and (2, 0)

Name _____ **Date** _____

Points in the Coordinate Plane:
Lesson 3B

Practice

Plot the given points on the coordinate grid. Use a ruler and draw a line that goes through the origin and the point you have placed on the grid.

1. (−2, 3)

3. (4, 1)

2. (−3, 0)

4. (−2, −2)

Plot the given points on the coordinate grid. Use a ruler and draw a line that goes through the two points you have placed on the grid.

5. (1, 4) and (2, −4)

7. (−2, 4) and (0, 1)

6. (3, 4) and (0, 4)

8. (−1, −4) and (3, 0)

Name _____ **Date** _____

Points in the Coordinate Plane:
Lesson 4A

Practice

1. Graph these points on the coordinate plane: A (5, 7); B (1, 7); C (1, 1); and D (5, 1). Then connect the points in alphabetical order. When you reach point D, stop. Identify the shape you have made.

The shape is a __letter C__

2. Graph these points on the coordinate plane: A (5, 7); B (1, 7); C (1, 1); D (5, 1); E (3, 4); and F (1, 4). Connect points A, B, C and D in alphabetical order. Stop and pick up your pencil. Now connect point E to point F. Stop. Identify the shape you have made.

The shape is a __letter E__

3. Graph these points on the coordinate plane: A (5, 1); B (5, 3); C (4, 4); D (5, 5); E (5, 7); F (1, 7); G (1, 4); H (1, 1). Connect points A, B, C, D, E, F, G, and H in alphabetical order. Connect point G to point C. Connect point A to point H. Stop. Identify the shape you have made.

The shape is a __letter B__

4. Graph these points on the coordinate plane: A (1, 1); B (3, 7); C (5, 1); D (2, 4); and E (4, 4). Connect points A, B, and C in alphabetical order. Stop and pick up your pencil. Now connect point D to point E. Stop. Identify the shape you have made.

The shape is a __letter A__

5. Graph these points on the coordinate plane: A (5, 1); B (5, 7); C (1, 1); D (1, 7); E (1, 4); and F (5, 4). Connect point A to point B. Connect point C to point D. Connect point F to point E. Stop. Identify the shape you have made.

The shape is a __letter H__

6. Rearrange your answers from Problems 1 through 5 to spell a very familiar word.

__B__ __E__ __A__ __C__ __H__

Name _____ **Date** _____

Points in the Coordinate Plane:
Lesson 4B

Practice

1. Graph these points on the coordinate plane: A (5, 7); B (1, 7); C (1, 1); and D (5, 1). Then connect the points in alphabetical order. When you reach point D, stop. Identify the shape you have made.

The shape is a __letter C__

2. Graph these points on the coordinate plane: A (5, 7); B (1, 7); C (1, 1); and D (5, 1). Then connect the points in alphabetical order. When you reach point D, connect D to point A. Stop. Identify the shape you have made.

The shape is a __letter O__

3. Graph these points on the coordinate plane: A (5, 7); B (5, 1); C (1, 1); and D (1, 7). Connect point A to point B. Connect point C to point D. Connect point D to point A. Stop. Identify the shape you have made.

The shape is a __letter N__

4. Graph these points on the coordinate plane: A (5, 7); B (1, 7); C (1, 1); D (5, 1); E(3,4); and F(1,4). Connect points A, B, C and D in alphabetical order. Stop and pick up your pencil. Now connect point E to point F. Stop. Identify the shape you have made.

The shape is a __letter E__

5. Graph these points on the coordinate plane: A (1, 1); B (3, 7); C (5, 1); D (2, 4); and E (4, 4). Connect points A, B, and C in alphabetical order. Stop and pick up your pencil. Now connect point D to point E. Stop. Identify the shape you have made.

The shape is a __letter A__

6. Rearrange your answers from Problems 1 through 5 to spell a very familiar word.

__O__ __C__ __E__ __A__ __N__

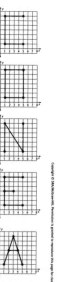

Line Properties: Lesson 1A

Practice

Determine the length of the line segments that connect these pairs of coordinates. Show your work.

1. (4, 2) and (8, 2)
   $8 - 4 = 4$

2. (−6, 2) and (0, 2)
   $0 - -6 = 6$

3. (0, 9) and (8, 9)
   $8 - 0 = 8$

4. (1, −7) and (12, −7)
   $12 - 1 = 11$

5. (−5, 3) and (−2, 3)
   $-2 - -5 = 3$

6. (−4, −3) and (−9, −3)
   $-4 - -9 = 5$

Determine the horizontal distance between the endpoints of the line segments that connect these pairs of coordinates. Show your work.

7. (5, 4) and (4, 2)
   $5 - 4 = 1$

8. (−4, 6) and (−9, 4)
   $-4 - -9 = 5$

9. (3, 9) and (−4, 6)
   $3 - -4 = 7$

10. (0, −7) and (6, −3)
    $6 - 0 = 6$

11. (0, 7) and (−5, 4)
    $0 - -5 = 5$

12. (4, −1) and (12, −7)
    $12 - 4 = 8$

Line Properties: Lesson 1B

Practice

Determine the length of the line segments that connect these pairs of coordinates. Show your work.

1. (5, 3) and (9, 3)
   $9 - 5 = 4$

2. (−3, 5) and (−5, 5)
   $-3 - -5 = 2$

3. (0, 6) and (3, 6)
   $3 - 0 = 3$

4. (8, 0) and (12, 0)
   $12 - 8 = 4$

5. (−7, 4) and (−2, 4)
   $-2 - -7 = 5$

6. (0, 12) and (−6, 12)
   $0 - -6 = 6$

Determine the horizontal distance between the endpoints of the line segments that connect these pairs of coordinates. Show your work.

7. (7, −2) and (5, −1)
   $7 - 5 = 2$

8. (−5, −8) and (−12, −3)
   $-5 - -12 = 7$

9. (−4, 6) and (9, −3)
   $9 - -4 = 13$

10. (0, 7) and (−6, 7)
    $0 - -6 = 6$

11. (2, 7) and (0, 4)
    $2 - 0 = 2$

12. (−2, −4) and (8, −6)
    $8 - -2 = 10$

Line Properties: Lesson 2A

Practice

Determine the length of the line segments that connect these pairs of coordinates. Show your work.

1. (4, 3) and (4, 7)
   $7 - 3 = 4$

2. (−3, 5) and (−3, 12)
   $12 - 5 = 7$

3. (0, 9) and (0, −7)
   $9 - -7 = 16$

4. (8, 0) and (8, 6)
   $6 - 0 = 6$

5. (−8, 6) and (−8, 3)
   $6 - 3 = 3$

6. (−6, −3) and (−6, 10)
   $10 - -3 = 13$

Determine the vertical distance between the endpoints of the line segments that connect these pairs of coordinates. Show your work.

7. (7, −2) and (5, −1)
   $-1 - -2 = 1$

8. (−5, −6) and (−10, −3)
   $-3 - -6 = 3$

9. (−4, 5) and (9, −3)
   $5 - -3 = 8$

10. (0, −4) and (−6, 7)
    $7 - -4 = 11$

11. (2, 7) and (3, 4)
    $7 - 4 = 3$

12. (−2, −4) and (8, −6)
    $-4 - -6 = 2$

Line Properties: Lesson 2B

Practice

Determine the length of the line segments that connect these pairs of coordinates. Show your work.

1. (5, 5) and (5, 12)
   $12 - 5 = 7$

2. (−3, 5) and (−3, 8)
   $8 - 5 = 3$

3. (0, 6) and (0, −4)
   $6 - -4 = 10$

4. (8, 0) and (8, −5)
   $0 - -5 = 5$

5. (−7, 4) and (−7, 9)
   $9 - 4 = 5$

6. (0, 12) and (0, 4)
   $12 - 4 = 8$

Determine the vertical distance between the endpoints of the line segments that connect these pairs of coordinates. Show your work.

7. (7, −2) and (5, −7)
   $-2 - -7 = 5$

8. (−2, −8) and (−12, −6)
   $-6 - -8 = 2$

9. (−4, 6) and (9, −8)
   $6 - -8 = 14$

10. (0, 7) and (−6, −3)
    $7 - -3 = 10$

11. (2, 10) and (0, 4)
    $10 - 4 = 6$

12. (−3, 5) and (8, 9)
    $9 - 5 = 4$

Practice

Using the right triangle on the coordinate grid to answer the following questions. You will also use the Pythagorean theorem to find the length of the Hypotenuse.

1. Find the length of side AB. Call it a in the Pythagorean theorem. Show your work.

   $8 - 3 = 5$, so $a = 5$

2. Square the length of side AB. Show your work.

   $5 \times 5 = 25$, so $a^2 = 25$

3. Find the length of side BC. Call it b in the Pythagorean theorem. Show your work.

   $13 - 1 = 12$, so $b = 12$

4. Square the length of side BC. Show your work.

   $12 \times 12 = 144$, so $b^2 = 144$

Use the Pythagorean theorem to find the length of the hypotenuse. Show your work below.

9. $a = 4$, $b = 3$, $c =$ ___5___

10 $a = 7$, $b = 24$, $c =$ ___25___

5. Write the Pythagorean theorem.

   $a^2 + b^2 = c^2$

6. Substitute the values for a^2 and b^2.

   $a^2 + b^2 = 25 + 144 = c^2$

7. Simplify the equation you have written by combining the numbers.

   $c^2 = 169$

8. What number multiplied by itself gives 169?

   13

Therefore, the length of the hypotenuse, side AC, in this right triangle is ___13___.

11. $a = 24$, $b = 10$, $c =$ ___26___

12. $a = 6$, $b = 8$, $c =$ ___10___

Practice

Use the right triangle on the coordinate grid to answer the following questions. You will also use the Pythagorean theorem to find the length of the hypotenuse.

1. Find the length of side KL. Call it a in the Pythagorean theorem. Show your work.

   $7 - 1 = 6$, so $a = 6$

2. Square the length of side KL. Show your work.

   $6 \times 6 = 36$, so $a^2 = 36$

3. Find the length of side LM. Call it b in the Pythagorean theorem. Show your work.

   $10 - 2 = 8$, so $b = 8$

4. Square the length of side LM. Show your work.

   $8 \times 8 = 64$, so $b^2 = 64$

5. Write the Pythagorean theorem.

   $a^2 + b^2 = c^2$

6. Substitute the values for a^2 and b^2.

   $a^2 + b^2 = 36 + 64 = c^2$

7. Simplify the equation you have written by combining the numbers.

   $c^2 = 100$

8. What number multiplied by itself gives 100?

   10

Therefore, the length of the hypotenuse, side KM, in this right triangle is ___10___.

Given the following endpoints, solve for the length of the line segment. You may find it helpful to plot the two points and the line segment between them on a piece of graph paper and draw the right triangle for which this line segment is the hypotenuse.

9. (1, 5) and (5, 8)

   The length is ___5___.

10. (2, 0) and (9, 24)

   The length is ___25___.

11. (2, 2) and (7, 14)

   The length is ___13___.

12. (2, 4) and (8, 12)

   The length is ___10___.

Practice

Consider these groups of three numbers and determine whether they are the sides in a right triangle or not. Explain your answer. You will use the Pythagorean theorem. In each case the hypotenuse will be identified by c.

1. 8 , 6, $c = 10$

   yes, because $8^2 + 6^2 = 10^2$,

   $64 + 36 = 100$

2. 4, 9, $c = 12$

   no, because $4^2 + 9^2 \neq 12^2$,

   $16 + 81 \neq 144$

3. 3, 6, $c = 7$

   no, because $3^2 + 6^2 \neq 7^2$,

   $9 + 36 \neq 49$

4. 4, 4, $c = 32$

   no, because $4^2 + 4^2 \neq 32^2$,

   $16 + 16 \neq 1024$

5. 6, 10, $c = 6$

   no, because $6^2 + 10^2 \neq 16^2$,

   $36 + 100 \neq 256$

6. 9, 8, $c = 12$

   no, because $9^2 + 8^2 \neq 12^2$,

   $81 + 64 \neq 144$

Find the missing leg or hypotenuse in each of these problems.

7. $a = 3$, $c = 5$, find b.

   $b =$ ___4___

8. $a = 12$, $b = 9$, find c.

   $c =$ ___15___

9. $b = 12$, $c = 13$, find a.

   $a =$ ___5___

10. $b = 15$, $c = 25$, find a.

   $a =$ ___20___

11. $a = 14$, $c = 50$, find b.

   $b =$ ___48___

12. $a = 10$, $b = 24$, find c.

   $c =$ ___26___

Practice

Consider these groups of three numbers and determine whether they are the sides in a right triangle or not. Explain your answer. You will use the Pythagorean theorem. In each case the hypotenuse will be identified by c.

1. 9, 12, $c = 15$

   yes, because $9^2 + 12^2 = 15^2$,

   $81 + 144 = 225$

2. 5, 10, $c = 12$

   no, because $5^2 + 10^2 \neq 12^2$,

   $25 + 100 \neq 144$

3. 4, 5, $c = 6$

   no, because $4^2 + 5^2 \neq 6^2$,

   $16 + 25 \neq 36$

4. 5, 12, $c = 17$

   no, because $5^2 + 12^2 \neq 17^2$,

   $25 + 144 \neq 289$

5. 0.6, 0.8, $c = 1$

   yes, because $0.6^2 + 0.8^2 = 1^2$,

   $0.36 + 0.64 = 1$

6. 3, 6, $c = 9$

   no, because $3^2 + 6^2 \neq 9^2$,

   $9 + 36 \neq 81$

Find the missing leg or hypotenuse in each of these problems. Use numbers you know to approximate the missing side length.

7. $a = 5$, $b = 6$, find c.

   $b =$ between ___7___ and ___8___

8. $a = 7$, $b = 9$, find c.

   $c =$ between ___11___ and ___12___

9. $b = 10$, $c = 12$, find a.

   $a =$ between ___6___ and ___7___

10. $b = 12$, $c = 15$, find a.

   $a =$ between ___7___ and ___8___

11. $a = 12$, $c = 20$, find b.

   $b =$ ___16___

12. $a = 3$, $b = 5$, find c.

   $c =$ between ___5___ and ___6___

Change and Slope: Lesson 1A

Practice

These two coordinate grids show lines with "special" slopes. Calculate the slopes and then answer the questions.

1.

The slope of this line is $\frac{0}{5}$ or ___0___

The slope of any horizontal line is ___0___

2.

The slope of this line is $\frac{4}{0}$ or ___undefined___

The slope of any vertical line is ___undefined___

Complete this table. Find the change in *y*-values and the change in the *x*-values for the pairs of points that are given. Then, calculate the slope.

	Points	Change in y-values	Change in x-values	Slope
3.	(2, 4) and (6, 5)	5 − 4 = 1	6 − 2 = 4	$\frac{1}{4}$
4.	(−2, −2) and (−4, −4)	−4 − −2 = −2	−4 − −2 = −2	$\frac{-2}{-2}$ or 1
5.	(3, 5) and (2, 7)	7 − 5 = 2	2 − 3 = −1	$\frac{2}{-1}$ or −2
6.	(4, −6) and (−5, 0)	0 − −6 = 6	−5 − 4 = −9	$\frac{6}{-9}$ or $\frac{2}{-3}$
7.	(5, 2) and (−5, −2)	−2 − 2 = −4	−5 − 5 = −10	$\frac{-4}{-10}$ or $\frac{2}{5}$
8.	(2, −2) and (−4, 4)	4 − −2 = 6	−4 − 2 = −6	$\frac{6}{-6}$ or −1
9.	(0, 0) and (1, 3)	3 − 0 = 3	1 − 0 = 1	$\frac{3}{1}$ or 3
10.	(−2, 3) and (6, 0)	0 − 3 = −3	6 − −2 = 8	$\frac{-3}{8}$
11.	(4, 6) and (0, 0)	0 − 6 = −6	0 − 4 = −4	$\frac{-6}{-4}$ or $\frac{3}{2}$
12.	(3, −4) and (1, −4)	−4 − −4 = 0	1 − 3 = −2	$\frac{0}{-2}$ or 0

Change and Slope: Lesson 1B

Practice

These two coordinate grids show lines with special slopes. Calculate the slopes and then answer the questions.

1.

The slope of this line is $\frac{6}{0}$ or ___undefined___

The slope of any vertical line is ___undefined___

2.

The slope of this line is $\frac{0}{-6}$ or ___0___

The slope of any horizontal line is ___0___

Complete this table. Find the change in *y*-values and the change in the *x*-values for the pairs of points that are given. Then, calculate the slope.

	Points	Change in y-values	Change in x-values	Slope
3.	(−2, 4) and (3, 5)	5 − 4 = 1	3 − −2 = 5	$\frac{1}{5}$
4.	(−2, −3) and (−4, −5)	−5 − −3 = −2	−4 − −2 = −2	$\frac{-2}{-2}$ or 1
5.	(0, 5) and (−2, 7)	7 − 5 = 2	−2 − 0 = −2	$\frac{2}{-2}$ or −1
6.	(−3, 5) and (−4, 0)	0 − 5 = −5	−4 − −3 = −1	$\frac{-5}{-1}$ or 5
7.	(4, 2) and (8, 4)	4 − 2 = 2	8 − 4 = 4	$\frac{2}{4}$ or $\frac{1}{2}$
8.	(2, 3) and (−5, 4)	4 − 3 = 1	−5 − 2 = −7	$\frac{1}{-7}$ or $-\frac{1}{7}$
9.	(0, 0) and (2, 8)	8 − 0 = 8	2 − 0 = 2	$\frac{8}{2}$ or 4
10.	(−1, 3) and (5, 0)	0 − 3 = −3	5 − −1 = 6	$\frac{-3}{6}$ or $-\frac{1}{2}$
11.	(5, 12) and (0, 0)	0 − 12 = −12	0 − 5 = −5	$\frac{-12}{-5}$ or $\frac{12}{5}$
12.	(7, −6) and (1, −4)	−4 − −6 = 2	1 − 7 = −6	$\frac{2}{-6}$ or $-\frac{2}{6}$

Change and Slope: Lesson 2A

Practice

Find the slopes using the points given on the graphs. Remember that slope is the change in the *y*-values written over the change in the *x*-values ("rise over run").

1.

The slope of this line is $\frac{1}{-7}$ or $\frac{-1}{7}$

2.

The slope of this line is $\frac{-6}{0}$ or undefined

3.

The slope of this line is $\frac{6}{6}$ or 1

4.

The slope of this line is $\frac{-4}{-8}$ or $\frac{1}{2}$

Use the formula to calculate the slope of the line between the given points. Compare your answers to those above.

5. (3, 2) and (−3, 8)
The slope of this line is $\frac{-6}{6}$ or −1

6. (1, 3) and (−2, −4)
The slope of this line is $\frac{-7}{-3}$ or $\frac{7}{3}$

7. (9, 5) and (9, 2)
The slope of this line is $\frac{-3}{0}$ or undefined

8. (4, 1) and (7, −9)
The slope of this line is $\frac{-10}{3}$

Change and Slope: Lesson 2B

Practice

Find each slope using the points given on the graphs. Remember that slope is the change in the *y*-values written over the change in the *x*-values ("rise over run").

1.

The slope of this line ___ $\frac{2}{-8}$ or $\frac{-1}{4}$

3.

The slope of this line is $\frac{4}{4}$ or 1

2.

The slope of this line is $\frac{4}{6}$ or $\frac{2}{3}$

4.

The slope of this line is $\frac{-3}{-2}$ or $\frac{3}{2}$

Use the formula to calculate the slope of the line between the given points. Compare your answers to those above.

5. (1, 1) and (−5, 8)
The slope of this line is $\frac{7}{-6}$ or $\frac{-7}{6}$

6. (−7, −6) and (2, 0)
The slope of this line is $\frac{6}{9}$ or $\frac{2}{3}$

7. (9, 3) and (5, 3)
The slope of this line is $\frac{0}{-4}$ or 0

8. (6, 4) and (1, 0)
The slope of this line is $\frac{-4}{5}$

Algebra Readiness • Practice Answers   237

Name _____ Date _____

Change and Slope: Lesson 3A

Practice

Graph the direct variation relationship 1 gallon = 4 quarts. Answer the questions to help you decide how the graph should look. Remember that two points determine a line. You will find a third point to make sure the other two points are correct.

1. Write two equations in x and y expressing this relationship.
   $y = 4x$ or $x = \frac{y}{4}$

2. Write three ordered pairs that satisfy this relationship.
   possible answers are
   (1, 4) , (2, 8) , and (3, 12)

3. Which quantity will you graph on the x-axis? ___gallons___

4. Which quantity will you graph on the y-axis? ___quarts___

5. What scale could you use on the x-axis? ___0 through 5___

6. What scale could you use on the y-axis? ___0 through 20___

7. Graph the relationship. **draw a line through (0, 0) and (2, 8) but do not put the coordinates on the graph**

8. Looking at the line on the graph, what is the x-value when $y = 16$? ___$x = 4$___

Here is the graph of the direct variation $y = -2x$. Looking at the graph, answer the following questions.

9. When $x = 2$, what is the value of y? ___-4___

10. When $y = 12$, what is the value of x? ___-6___

11. When $y = 0$, what is the value of x? ___0___

12. When $x = -6$, what is the value of y? ___12___

13. When $x = 5$, what is the value of y? ___-10___

14. When $y = -8$, what is the value of x? ___4___

Algebra Readiness • Practice   157

Name _____ Date _____

Change and Slope: Lesson 3B

Practice

Graph the direct variation relationship 1 cup = 8 ounces. Answer the questions to help you decide how the graph should look. Remember that two points determine a line. You will find a third point to make sure the other two points are correct.

1. Write two equations in x and y expressing this relationship.
   $y = 8x$ or $x = \frac{y}{8}$

2. Write three ordered pairs that satisfy this relationship.
   possible answers are
   (1, 8) , (2, 16) and (3, 24)

3. Which quantity will you graph on the x-axis? ___cups___

4. Which quantity will you graph on the y-axis? ___ounces___

5. What scale could you use on the x-axis? ___0 through 5___

6. What scale could you use on the y-axis? ___0 through 40, by twos___

7. Graph the relationship.
   draw a line through (0, 0) and (2, 16) but do not put the coordinates on the graph

8. Looking at the line on the graph, what is the x-value when $y = 16$? ___$x = 2$___

Here is the graph of the direct variation $y = 5x$. Looking at the graph, answer the following questions.

9. When $x = 2$, what is the value of y? ___10___

10. When $y = -15$, what is the value of x? ___-3___

11. When $y = 0$, what is the value of x? ___0___

12. When $x = -2$, what is the value of y? ___-10___

13. When $x = 4$, what is the value of y? ___20___

14. When $y = -5$, what is the value of x? ___-1___

158   Algebra Readiness • Practice

Name _____ Date _____

Change and Slope: Lesson 4A

Practice

The perimeter of a regular octagon, like a stop sign, can be found by adding the lengths of the 8 equal sides or by using the formula $P = 8s$, where s is the length of one of the sides. Answer the following questions and then graph this relationship.

1. Write the formula as an equation involving x and y. ___$y = 8x$ or $x = \frac{y}{8}$___

2. Would you pick negative values for x or y? Why or why not?
   ___Possible answer: No, because length is measured in positive values.___

3. Write three ordered pairs of numbers that can help you graph this line. Remember to keep the x-values fairly small so that the graph doesn't get too large. ___possible answers (0, 0), (1, 8), (2, 16), (3, 24)___

4. What quantity will be indicated on the horizontal axes? ___the length of a side___

5. What quantity will be indicated on the vertical axes? ___the perimeter___

6. Plot the three points that you picked and draw the line through those points.

7. From the points you graphed on the coordinate grid, calculate the slope of the line (change in y-values/change in x-values or rise/run).
   Possible answer: $\frac{(24-16)}{(3-2)} = \frac{8}{1} = 8$

For each pair of points below, calculate the slope (ratio) of the line between them. You may want to plot the points on a graph to help you find the slope.

8. (3, 5) and (4, 1)
   -4

9. $(-2, -6)$ and $(3, -2)$
   $\frac{4}{5}$

10. (0, 4) and (5, -2)
    $-\frac{6}{5}$

11. (3, 2) and $(-5, 0)$
    $\frac{-2}{-8}$ or $\frac{1}{4}$

Algebra Readiness • Practice   159

Name _____ Date _____

Change and Slope: Lesson 4B

Practice

The number of quarters in a given number of dollars can be described by the formula $Q = 4D$, where D is the number of dollars. Answer the following questions and then graph this relationship.

1. Write the formula as an equation involving x and y. ___$y = 4x$ or $x = \frac{y}{4}$___

2. Would you pick negative values for x or y? Why or why not?
   ___Possible answer: No, because the number of quarters is positive.___

3. Write three ordered pairs of numbers that can help you graph this line. Remember to keep the x values fairly small so that the graph doesn't get too large. ___possible answers (0, 0), (1, 4), (2, 8), (3, 12)___

4. What quantity will be indicated on the horizontal axes? ___the number of dollars___

5. What quantity will be indicated on the vertical axes? ___the number of quarters___

6. Plot the three points that you picked and draw the line through those points.

7. From the points you graphed on the coordinate grid, calculate the slope of the line (change in y-values/change in x-values or rise/run).
   possible answer: $\frac{(12-8)}{(3-2)} = \frac{4}{1} = 4$

For each pair of points below, calculate the slope (ratio) of the line between them. You may want to plot the points on a graph to help you find the slope.

8. (5, 2) and (2, 7)
   $\frac{5}{3}$

9. $(-1, 5)$ and $(2, -4)$
   $-\frac{9}{3}$ or -3

10. (6, 0) and $(3, -1)$
    $\frac{1}{3}$

11. (6, 1) and $(0, -5)$
    $\frac{-6}{-6}$ or 1

160   Algebra Readiness • Practice

Rates and Products: Lesson 1A

Practice

Consider the equation $y = -2x + 4$. Answer the following questions, and then graph the line that corresponds to this equation.

1. What is the y-intercept of this line? ___4___

2. Write the coordinates of the y-intercept. ___(0, 4)___

3. What is the slope of this line? ___-2 or $-\frac{2}{1}$___

4. From the y-intercept the slope tells you to move down ___2___ units and then to the ___right___ 1 unit to find another point on this line. The coordinates of this point are ___(1, 2)___.

5. From the y-intercept you could also move up ___2___ units and then back to the left ___1___ unit to find a third point on this line. The coordinates of this point are ___(−1, 6)___.

6. Graph the line which corresponds to the equation $y = -2x + 4$.

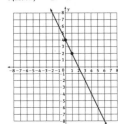

7. Following the same process that you used to graph $y = -2x + 4$, graph the line that corresponds to the equation $y = \frac{2}{3}x - 1$.

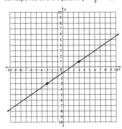

Rates and Products: Lesson 1B

Practice

Consider the equation $y = -3x + 1$. Answer the following questions, and then graph the line that corresponds to this equation.

1. What is the y-intercept of this line? ___1___

2. Write the coordinates of the y-intercept. ___(0, 1)___

3. What is the slope of this line? ___-3 or $-\frac{3}{1}$___

4. From the y-intercept the slope tells you to move down ___3___ units and then to the ___right___ 1 unit to find another point on this line. The coordinates of this point are ___(1, −2)___.

5. From the y-intercept you could also move up ___3___ units and then back to the left ___1___ unit to find a third point on this line. The coordinates of this point are ___(−1, 4)___.

6. Graph the line which corresponds to the equation $y = -3x + 1$.

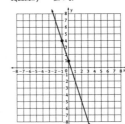

7. Following the same process that you used to graph $y = -3x + 1$, graph the line that corresponds to the equation $y = \frac{1}{4}x + 2$.

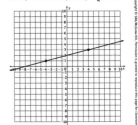

Rates and Products: Lesson 2A

Practice

The information contained in this chart shows the fuel efficiency for three cars used in a recent test. Find the rate (miles per gallon) for each car and then decide which car is the most fuel-efficient. (Hint: When finding miles per gallon in a test, you usually round to one decimal place in order to see a difference in the rate.)

Fuel Efficiency Test Results

	Vehicle	Miles	Gallons	Miles per Gallon (unit rate)
1.	Car 1	250	8	31.3
2.	Car 2	240	7.6	31.6
3.	Car 3	248	8	31

4. How would you write the rate per gallon for Car 3 rounded to the nearest tenth so you could compare it to the other two rates? ___31.0___

5. Which vehicle has the best fuel efficiency? ___Car 2___

6. Which vehicle has the worst fuel efficiency? ___Car 3___

The information contained in this chart shows prices for three brands of trail mix in a local market. Find the rate (cost per ounce) for each brand and then decide which brand has the best price for trail mix. (Hint: When finding unit cost, you usually round to three decimal places in order to see a difference in the unit cost.)

Trail Mix Comparison

	Brand	Ounces	Price per Package	Cost per Ounce (unit rate)
7.	Brand A	12	$1.86	$0.155
8.	Brand B	15	$2.19	$0.146
9.	Brand C	20	$2.95	$0.148

10. Describe the process you use to find the cost per ounce. ___Divide the cost by the number of ounces.___

11. Which brand has the best price per ounce? ___Brand B___

12. Which brand is the most expensive per ounce? ___Brand A___

Rates and Products: Lesson 2B

Practice

A small packaging factory has three assembly lines that package sport cards for collectors. They package cards about football players, soccer players, baseball players, hockey players, and others. The information contained in this chart shows the production rate for the three lines. Find the rate (packages per hour) for each line and complete the information in the chart.

Assembly Line Efficiency Results

	Line	Packages	Hours	Packages per Hour (unit rate)
1.	Line 1	183	3	61
2.	Line 2	145	2.5	58
3.	Line 3	165	2.75	60

4. Which Line is the most efficient? ___Line 1___

5. Which Line is the least efficient? ___Line 2___

6. At this rate, how many packages would Line 1 produce in an 8-hour day? ___488 packages___

7. At this rate, how many packages would Line 2 produce in an 8-hour day? ___464 packages___

8. At this rate, how many packages would Line 3 produce in an 8-hour day? ___480 packages___

Algebra Readiness • Practice Answers   239

Practice

Ki and Miguel have a small roofing company. They often get jobs working as subcontractors for companies who are building houses in subdivisions. Their team of 5 roofers can usually complete the roofs on 3 houses in 6 days.

1. How many person-days are needed for this 3-house project?
   __30 person-days__

2. At this rate, how long would it take one roofer to complete the job if he or she had to work all alone? __30 days__

3. At this rate, how much of the project does each roofer complete in one day? __$\frac{1}{30}$__

4. How many days would it take if Ki and Miguel hired 6 roofers? __5 days__

5. How many days would it take if they hired 10 roofers? __3 days__

6. How many roofers would they need to hire to complete this job in 2 days? __15 roofers__

Tony and Betsy do the landscaping for this same developer. The two of them working together can finish the landscaping for a house in 10 hours.

7. How many person-hours are needed for a one-house project? __20 person-hours__

8. At this rate, how long would it take one of them to complete the job if he or she had to work all alone? __20 hours__

9. At this rate, how much of the project does each of them complete in one hour? __$\frac{1}{20}$__

10. How many hours would it take if Betsy and Tony hired another worker? __$6\frac{2}{3}$ hours__

11. How many hours would it take if they hired 2 more workers? __5 hours__

12. How many workers would they need to hire to complete this job in 4 hours? __3 more workers__

Practice

Perry and Tammy have a small painting company. They often get jobs working as subcontractors for companies who are building houses in subdivisions. Their team of 4 painters can usually complete the inside painting on 2 houses in 5 days.

1. How many person-days are needed for this 2-house project? __20 person-days__

2. At this rate, how long would it take one painter to complete the job if he or she had to work all alone? __20 days__

3. At this rate, how much of the project does each painter complete in one day? __$\frac{1}{20}$__

4. How many days would it take if Perry and Tammy hired 5 painters? __4 days__

5. How many days would it take if they only had 2 painters? __10 days__

6. How many painters would they need to hire to complete this 2-house project in 2 days? __10 painters__

Luis is a subcontractor who clears the land for this same developer. Luis's crew of 3 workers can clear the land for 2 houses in 8 hours.

7. How many person-hours are needed for a 2-house clearing? __24 person-hours__

8. At this rate, how long would it take one of them to complete the job if he or she had to work all alone? __24 hours__

9. At this rate, how much of the project does each of them complete in one hour? __$\frac{1}{24}$__

10. How many hours would it take if Luis hired another worker? __8 hours__

11. How many hours would it take if he hired 3 more workers? __4 hours__

12. How many workers would he need to use to complete this job in 6 hours? __4 workers__

Practice

The average speed at a recent Indianapolis 500 Mile Race was 206 miles per hour. At this rate, how many feet per minute does a racecar travel?

1. Write 206 miles per hour as a ratio (in fraction form) but keep the units with the numbers.
$$\frac{206 \text{ miles}}{1 \text{ hour}}$$

2. Make a multiplication expression using all 3 ratios. Be sure to include the labels with the numbers.
$$\frac{206 \text{ miles}}{1 \text{ hour}} \times \frac{5{,}280 \text{ feet}}{1 \text{ mile}} \times \frac{1 \text{ hour}}{60 \text{ minutes}}$$

3. Use Dimensional Analysis to cancel out the units. Since the original question was feet per minute, you want feet remaining in the top and minutes in the bottom.
$$\frac{206 \text{ miles}}{1 \text{ hour}} \times \frac{5{,}280 \text{ feet}}{1 \text{ mile}} \times \frac{1 \text{ hour}}{60 \text{ minutes}}$$

4. Now finish the mathematics by multiplying across the numerators and denominators, reducing first as you go, or reduce the final result.
$$\frac{206 \times 5280 \times 1 \text{ feet}}{1 \times 1 \times 60 \text{ minutes}} = \frac{1{,}087{,}680 \text{ feet}}{60 \text{ minutes}} = 18{,}128 \text{ feet per minute}$$

5. Finally, answer the question using the correct labels.
   These race cars were traveling 18,128 feet per minute.

Find the error in this solution. Then, give the correct answer.

6. Change 150 feet per second to miles per hour.
$$\frac{150 \text{ ft}}{1 \text{ sec}} \times \frac{60 \text{ sec}}{1 \text{ hour}} \times \frac{1 \text{ mi}}{5{,}280 \text{ ft}} = \frac{150 \times 60 \times 1}{1 \times 1 \times 5{,}280}$$
$$= \frac{9000}{5280} = 1.705 \text{ miles per hour}$$
They need to change seconds to minutes and then minutes to hours.

The correct answer is __102 miles per hour__

Practice

Hermani, a major league baseball pitcher, throws his fastball at an average rate of 94 miles per hour. How fast would that pitch be in feet per second?

1. Write 94 miles per hour as a ratio (in fraction form) but keep the units with the numbers.
$$\frac{94 \text{ miles}}{1 \text{ hour}}$$

2. Make a multiplication expression using all 4 ratios. Be sure to include the labels with the numbers.
$$\frac{94 \text{ miles}}{1 \text{ hour}} \times \frac{5{,}280 \text{ feet}}{1 \text{ mile}} \times \frac{1 \text{ hour}}{60 \text{ minutes}} \times \frac{1 \text{ minute}}{60 \text{ seconds}}$$

3. Use Dimensional Analysis to cancel out the units. Since the original question was feet per second, you want feet remaining in the top and seconds in the bottom.
$$\frac{94 \text{ miles}}{1 \text{ hour}} \times \frac{5{,}280 \text{ feet}}{1 \text{ mile}} \times \frac{1 \text{ hour}}{60 \text{ minutes}} \times \frac{1 \text{ minute}}{60 \text{ seconds}}$$

4. Now finish the mathematics by multiplying across the numerators and denominators, reducing first as you go, or reduce the final result.
$$\frac{94 \times 5280 \times 1 \times 1 \text{ feet}}{1 \times 1 \times 60 \times 60 \text{ minutes}} = \frac{496{,}320 \text{ feet}}{3600 \text{ seconds}} = 137.9 \text{ feet per second}$$

5. Finally, answer the question using the correct labels.
   Hermani's fastball was traveling 137.9 feet per second.

Find the error in this solution. Then, give the correct answer.

6. Convert $4.98 per quart to cost per ounce.
$$\frac{\$4.98}{1 \text{ quart}} \times \frac{1 \text{ quart}}{2 \text{ pints}} \times \frac{1 \text{ pint}}{2 \text{ cups}} \times \frac{1 \text{ cup}}{16 \text{ ounces}} = \frac{\$4.98 \times 1 \times 1 \times 1}{1 \times 2 \times 2 \times 16}$$
$$= \frac{4.98}{64} = \$0.078 \text{ per ounce}$$
Sample answer: There are 8 ounces in a cup, not 16.

The correct answer is __$0.156 per ounce__

240 Algebra Readiness • Practice Answers

Name _____ Date _____

Manipulating Equations: Lesson 1A

Practice

Write *opposite* or *reciprocal* on the blank to make each statement true.

1. 5 is the __reciprocal__ of $\frac{1}{5}$.   5. $\frac{1}{4}$ is the __reciprocal__ of 4.

2. -8 is the __opposite__ of 8.   6. $-\frac{1}{4}$ is the __opposite__ of $\frac{1}{4}$.

3. $\frac{2}{3}$ is the __opposite__ of $-\frac{2}{3}$.   7. $-\frac{5}{7}$ is the __reciprocal__ of $-\frac{7}{5}$.

4. $\frac{2}{3}$ is the __reciprocal__ of $\frac{3}{2}$.   8. 1 is the __reciprocal__ of 1.

	Number	Opposite	Reciprocal
9.	9	-9	$\frac{1}{9}$
10.	-6	6	$-\frac{1}{6}$
11.	$\frac{3}{5}$	$-\frac{3}{5}$	$\frac{5}{3}$
12.	$\frac{1}{2}$	$-\frac{1}{2}$	2
13.	$-\frac{5}{8}$	$\frac{5}{8}$	$-\frac{8}{5}$
14.	$\frac{1}{3}$	$-\frac{1}{3}$	3
15.	-1	1	-1

Name _____ Date _____

Manipulating Equations: Lesson 1B

Practice

Draw a solid line from the number to its opposite or a dashed line from the number to its reciprocal.

1. $\frac{5}{8}$ A. $\frac{1}{12}$
2. -6 B. $\frac{-5}{8}$
3. $\frac{1}{4}$ C. $\frac{-1}{5}$
4. $\frac{-2}{3}$ D. $\frac{2}{3}$
5. 12   E. $\frac{-1}{6}$
6. $\frac{1}{5}$ F. 4

	Number	Opposite	Reciprocal
7.	$\frac{1}{3}$	$-\frac{1}{3}$	3
8.	10	-10	$\frac{1}{10}$
9.	$\frac{3}{4}$	$-\frac{3}{4}$	$\frac{4}{3}$
10.	-7	7	$-\frac{1}{7}$
11.	$-\frac{3}{5}$	$\frac{3}{5}$	$-\frac{5}{3}$
12.	6	-6	$\frac{1}{6}$

Name _____ Date _____

Manipulating Equations: Lesson 2A

Practice

Draw a line from the expression in Column 1 to the simplified expression in Column 2 when the distributive property has been correctly applied.

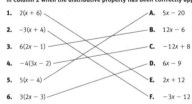

1. $2(x + 6)$ A. $5x - 20$
2. $-3(x + 4)$ B. $12x - 6$
3. $6(2x - 1)$ C. $-12x + 8$
4. $-4(3x - 2)$ D. $6x - 9$
5. $5(x - 4)$ E. $2x + 12$
6. $3(2x - 3)$ F. $-3x - 12$

7. Is $-2(x + 4) = -2x - 8$ a true or false statement? Prove whether this application of the distributive property is true or false by letting $x = 10$. Show your work. If the application is false, give the correct simplification.

$$-2(x + 4) = -2x - 8$$
$$-2(10 + 4) = -2(10) - 8$$
$$-2(14) = -20 - 8$$
$$-28 = -28$$
True

8. Is $9x - 6 \div 3 = 3x - 3$ a true or false statement? Prove whether this application of the distributive property is true or false by letting $x = 5$. Show your work. If the application is false, give the correct simplification.

$$(9x - 6) \div 3 = 3x - 3$$
$$[9(5) - 6] \div 3 = 3(5) - 3$$
$$(45 - 6) \div 3 = 15 - 3$$
$$39 \div 3 = 12$$
$$13 = 12$$
False. $(9x - 6) \div 3 = 3x - 2$

Name _____ Date _____

Manipulating Equations: Lesson 2B

Practice

1. Is $4(3x + 1) = 12x + 1$ a true or false statement? Prove whether this application of the distributive property is true or false by letting $x = 2$. Show your work. If the application is false, give the correct simplification.

$$4(3x + 1) = 12x + 1$$
$$4[3(2) + 1] = 12(2) + 1$$
$$4(6 + 1) = 24 + 1$$
$$4(7) = 25$$
$$28 = 25$$
False. $4(3x + 1) = 12x + 3$

2. Is $-2(x - 4) = -2x + 8$ a true or false statement? Prove whether this application of the distributive property is true or false by letting $x = 6$. Show your work. If the application is false, give the correct simplification.

$$-2(x - 4) = -2x + 8$$
$$-2(6 - 4) = -2(6) + 8$$
$$-2(2) = -12 + 8$$
$$-4 = -4$$
True

Simplify each of the following expressions by correctly applying the distributive property.

3. $6(x - 7)$ __$6x - 42$__   7. $-4(2x + 1)$ __$-8x - 4$__

4. $(-9x - 18) \div 9$ __$-x - 2$__   8. $-5(7 + 3x)$ __$-35 - 15x$__

5. $(8x - 24) \div 4$ __$2x - 6$__   9. $(15x - 20) \div 5$ __$3x - 4$__

6. $7(-3x + 2)$ __$-21x + 14$__   10. $(-12x + 6) \div -2$ __$6x - 3$__

Algebra Readiness • Practice Answers   **241**

Practice

Write the inverse of each given equation.

1. If $\sqrt{49} = 7$, then ____$7^2 = 49$____.

2. If $6^4 = 1,296$, then ____$\sqrt[4]{1,296} = 6$____.

3. If $3^5 = 243$, then ____$\sqrt[5]{243} = 3$____.

4. If $\sqrt[3]{64} = 4$, then ____$4^3 = 64$____.

5. If $9^2 = 81$, then ____$\sqrt{81} = 9$____.

6. If $\sqrt[4]{10,000} = 10$, then ____$10^4 = 10,000$____.

Use your calculator to find the roots of these numbers. Round your answers to the nearest thousandth.

7. $\sqrt[3]{12} =$ ___2.289___

8. $\sqrt[4]{1,000} =$ ___5.623___

9. $\sqrt[4]{85} =$ ___2.432___

10. $\sqrt{29} =$ ___5.385___

Simplify each variable expression.

11. $\sqrt[3]{(64x^3)} =$ ___$4x$___

12. $\sqrt[4]{(81x^8)} =$ ___$3x^2$___

Practice

Write the inverse of each given equation.

1. If $\sqrt{81} = 9$, then ____$9^2 = 81$____.

2. If $5^3 = 125$, then ____$\sqrt[3]{125} = 5$____.

3. If $2^8 = 256$, then ____$\sqrt[8]{256} = 2$____.

4. If $\sqrt[4]{1,296} = 6$, then ____$6^4 = 1,296$____.

5. If $12^3 = 1,728$, then ____$\sqrt[3]{1,728} = 12$____.

6. If $\sqrt[5]{100,000} = 10$, then ____$10^5 = 100,000$____.

Use your calculator to find the roots of these numbers. Round your answers to the nearest thousandth.

7. $\sqrt[4]{100} =$ ___3.162___

8. $\sqrt[3]{13} =$ ___3.606___

9. $\sqrt[3]{30} =$ ___3.107___

10. $\sqrt[5]{250} =$ ___3.017___

Simplify each variable expression.

11. $\sqrt[6]{(x^{12})} =$ ___x^2___

12. $\sqrt[3]{(125x^9)} =$ ___$5x^3$___

Practice

Draw a line from the power expression in Column 1 to its simplification in Column 2.

Column 1		Column 2	
1. 7^2	D	**A.** $6x$	
2. $64^{\frac{1}{2}}$	J	**B.** $16x^4$	
3. 6^3	H	**C.** $32x^5$	
4. $125^{\frac{1}{3}}$	I	**D.** 49	
5. $(-3x)^3$	G	**E.** $7x^2y^2$	
6. $(2x)^5$	C	**F.** $121x^2$	
7. $(36x^2)^{\frac{1}{2}}$	A	**G.** $-27x^3$	
8. $(49x^4y^4)^{\frac{1}{2}}$	E	**H.** 216	
9. $(-2x)^4$	B	**I.** 5	
10. $(11x)^2$	F	**J.** 8	

Practice

Draw a line from the power expression in Column 1 to its simplification in Column 2.

Column 1		Column 2	
1. 4^3	D	**A.** $2xy^2$	
2. $4^{\frac{1}{2}}$	G	**B.** $100x^4$	
3. 8^3	E	**C.** $-125x^3$	
4. $64^{\frac{1}{3}}$	H	**D.** 64	
5. $(-5x)^3$	C	**E.** 512	
6. $(3x)^2$	I	**F.** $256x^4$	
7. $(64x^4)^{\frac{1}{2}}$	J	**G.** 2	
8. $(8x^3y^6)^{\frac{1}{3}}$	A	**H.** 4	
9. $(-4x)^4$	F	**I.** $9x^2$	
10. $(10x^2)^2$	B	**J.** $8x^2$	

Exponents and Applications: Lesson 1A

Practice

Use the Pythagorean theorem to determine whether each group of three numbers form a right triangle. Explain your answer. The hypotenuse equals *c*.

1. 8, 6, $c = 10$

   yes, because $8^2 + 6^2 = 10^2$, $64 + 36 = 100$

2. 2, 4, $c = 5$

   no, because $2^2 + 4^2 \neq 5^2$, $4 + 16 \neq 25$

3. $\sqrt{10}$, 4, $c = \sqrt{26}$

   yes, because $\sqrt{10}^2 + 4^2 = \sqrt{26}^2$, $10 + 16 = 26$

4. 10, 24, $c = 26$

   yes, because $10^2 + 24^2 = 26^2$, $100 + 576 = 676$

Find the missing leg or hypotenuse. If your answers are not perfect squares, leave the answers in square root form.

5. $a = 12$, $b = 5$, find *c*.
   13

6. $a = 36$, $c = 60$, find *b*.
   48

7. $b = 9$, $c = 15$, find *a*.
   12

8. $b = \sqrt{8}$, $c = \sqrt{24}$, find *a*.
   4

9. $a = 5$, $c = 9$, find *b*.
   $\sqrt{56}$

10. $a = 24$, $c = 25$, find *b*.
    7

Exponents and Applications: Lesson 1B

Practice

Use the Pythagorean theorem to determine whether each group of three numbers form a right triangle. Explain your answer. The hypotenuse equals *c*.

1. 16, 12, $c = 20$

   yes, because $16^2 + 12^2 = 20^2$, $256 + 144 = 400$

2. 4, 8, $c = 10$

   no, because $4^2 + 8^2 \neq 10^2$, $16 + 64 \neq 100$

3. $\sqrt{8}$, 3, $c = \sqrt{17}$

   yes, because $\sqrt{8}^2 + 3^2 = \sqrt{17}^2$, $8 + 9 = 17$

4. 7, 4, $c = 8$

   no, because $7^2 + 4^2 \neq 8^2$, $49 + 16 \neq 64$

Find the missing leg or hypotenuse. If your answers are not perfect squares, leave the answers in square root form.

5. $a = 8$, $b = 6$, find *c*.
   10

6. $a = 6$, $c = 9$, find *b*.
   $\sqrt{45}$

7. $b = 10$, $c = 26$, find *a*.
   24

8. $b = \sqrt{11}$, $c = 6$, find *a*.
   5

9. $a = 21$, $c = 75$, find *b*.
   72

10. $a = 24$, $b = 18$, find *c*.
    30

Exponents and Applications: Lesson 2A

Practice

State whether the following statements are *true* or *false*. If the statement is false, write the correct answer.

1. $3^2 = 9$
   True

2. $3^{-2} = \frac{1}{9}$
   True

3. $4^3 = 12$
   False, $4^3 = 64$

4. $4^{-1} = -4$
   False, $4^{-1} = \frac{1}{4}$

5. $4^0 = 4$
   False, $4^0 = 1$

6. $3^1 = 3$
   True

7. $3^4 = 81$
   True

8. $4^{-3} = -12$
   False, $4^{-3} = \frac{1}{64}$

Simplify. Assume all variables do not equal 0.

9. $(2x)^1 = $ ___$2x$___

10. $(4x)^3 = $ ___$64x^3$___

11. $(-4x)^0 = $ ___1___

12. $(4x^3)^{-1} = $ ___$\frac{1}{4x^3}$___

13. $(6x)^2 = $ ___$36x^2$___

14. $(-2xy^2)^0 = $ ___1___

Exponents and Applications: Lesson 2B

Practice

State whether the following statements are *true* or *false*. If the statement is false, write the correct answer.

1. $7^3 = 343$
   True

2. $1^{-7} = -7$
   False, $1^7 = \frac{1}{1^7} = 1$

3. $7^{-3} = -343$
   False, $7^{-3} = \frac{1}{343}$

4. $7^{-1} = \frac{1}{7}$
   True

5. $7^0 = 7$
   False, $7^0 = 1$

6. $7^4 = 2,401$
   True

7. $7^{-2} = \frac{1}{49}$
   True

8. $7^2 = 14$
   False, $7^2 = 49$

Simplify. Assume all variables do not equal 0.

9. $(-3x)^2 = $ ___$9x^2$___

10. $(-x)^{-2} = $ ___$\frac{1}{x^2}$___

11. $(5xy)^0 = $ ___1___

12. $(-5x)^1 = $ ___$-5x$___

13. $(3xy)^{-1} = $ ___$\frac{1}{3xy}$___

14. $(6xy)^0 = $ ___1___

Algebra Readiness • Practice Answers   243

Lesson 3A

Chapter 22

Name _____ **Date** _____

Exponents and Applications: Lesson 3A

Practice

State whether the following statements are *true* or *false*. If the statement is false, write the correct answer. Leave the answers in exponential form.

1. $4^3 \times 4^4 = 4^7$
   _____ True _____

2. $(4x)^3 = 64x^3$
   _____ True _____

3. $3^8 \div 3^4 = 3^2$
   False, $3^8 \div 3^4 = 3^4$

4. $(3x)^4 \times (3x)^2 = (3x)^8$
   False, $(3x)^4 \times (3x)^2 = (3x)^6$

5. $4^2 \times 4^6 = 4^{12}$
   False, $4^2 \times 4^6 = 4^8$

6. $(4xy)^6 \div (4xy)^3 = (4xy)^3$
   _____ True _____

7. $(3^4)^3 = 3^{12}$
   _____ True _____

8. $(3x^2)^3 = 9x^6$
   False, $(3x^2)^3 = 27x^6$

9. $\left(\frac{3}{4}\right)^3 = \frac{27}{64}$
   _____ True _____

10. $\left(\frac{x}{4}\right)^2 = \frac{x^2}{16}$
   _____ True _____

11. $4^6 \div 4^3 = 4^2$
   False, $4^6 \div 4^3 = 4^3$

12. $(4xy)^3 \times (4xy)^4 = (4xy)^{12}$
   False, $(4xy)^3 \times (4xy)^4 = 4xy^7$

Simplify. Assume all variables do not equal 0.

13. $(-3x)^0 = $ _____ 1 _____

14. $(4x)^3 \times (4x)^3 = $ _____ $(4x)^6$, or $4{,}096x^6$ _____

15. $(4xy)^2 = $ _____ $16x^2y^2$ _____

16. $(3xy)^5 \div (3xy)^3 = $ _____ $(3xy)^2$, or $9x^2y^2$ _____

17. $(-3xy)^1 = $ _____ $-3xy$ _____

18. $\left(\frac{3}{xy}\right)^3 = $ _____ $\frac{27}{x^3y^3}$ _____

Algebra Readiness • Practice **181**

Lesson 3B

Chapter 22

Name _____ **Date** _____

Exponents and Applications: Lesson 3B

Practice

State whether the following statements are *true* or *false*. If the statement is false, write the correct answer. Leave the answers in exponential form. Assume all variables are not equal to zero.

1. $5^3 \times 5^2 = 5^5$
   _____ True _____

2. $(5x)^2 = 25x^2$
   _____ True _____

3. $5^{10} \div 5^2 = 5^5$
   False, $5^{10} \div 5^2 = 5^8$

4. $(-3x)^4 = -81x^4$
   False, $(-3x)^4 = 81x^4$

5. $(2^3)^2 = 2^5$
   False, $(2^3)^2 = 2^6$

6. $(8xy)^6 \div (8xy)^5 = (8xy)^1$, or $8xy$
   _____ True _____

7. $\left(\frac{5}{2}\right)^4 = 5^{\frac{4}{2}}$
   False, $\left(\frac{5}{2}\right)^4 = \frac{5^4}{2^4}$

8. $(5x^2) \times (5x^2)^3 = (5x^2)^4 = 5x^8$
   False, $(5x^2) \times (5x^2)^3 = (5x^2)^4 = 625x^8$

9. $5^4 \div 5^1 = 5^3$
   _____ True _____

10. $\left(\frac{x^2}{5}\right)^2 = \frac{x^4}{5^2}$, or $\frac{x^4}{25}$
    _____ True _____

11. $2^4 \times 2^4 = 4^4$
    False, $2^4 \times 2^4 = 2^8$

12. $(-2xy)^5 \times (-2xy)^4 = (-2xy)^{20}$
    False, $(-2xy)^5 \times (-2xy)^4 = (-2xy)^9$

Simplify. Assume all variables do not equal 0.

13. $(6x^2)^1 = $ _____ $6x^2$ _____

14. $\left(\frac{2xy}{3}\right)^2 = $ _____ $\frac{4x^2y^2}{9}$ _____

15. $(-2xy)^0 = $ _____ 1 _____

16. $(6xy)^4 \div (6xy)^3 = $ _____ $(6xy)^1$, or $6xy$ _____

17. $(6xy) \times (6xy)^2 = $ _____ $(6xy)^3$, or $216x^3y^3$ _____

18. $(2x^2)^4 = $ _____ $16x^8$ _____

182 Algebra Readiness • Practice

Lesson 4A

Chapter 22

Name _____ **Date** _____

Exponents and Applications: Lesson 4A

Practice

Each of the following simplifications has an error. Find the error. Explain what should have been done. Then simplify the expression following the 5 Basic Rules. Assume all variables are not equal to zero.

1. $x^5 \times x^5 = x^{25}$
   need to add exponents, not multiply them
   The correct simplification is _____ $x^5 \times x^5 = x^{10}$ _____

2. $(-4x^2)(2x^2) = -8x^2$
   need to multiply x^2 by x^2 also, not just keep it as it is
   The correct simplification is _____ $(-4x^2)(2x^2) = -8x^4$ _____

3. $\frac{x^{10}}{x^2} = x^5$
   need to subtract the exponents, not divide them
   The correct simplification is _____ $\frac{x^{10}}{x^2} = x^8$ _____

4. $(-3x^2)^4 = -3x^8$
   did not take -3 to the 4th power
   The correct simplification is _____ $(-3x^2)^4 = 81x^8$ _____

Simplify each expression using the 5 Basic Rules, PEMDAS, the distributive property, and the rules for signed numbers. Assume all variables do not equal 0.

5. $x^3 \times x^6 = $ _____ x^9 _____

6. $-4x^2(-3x+1) = $ _____ $12x^3 - 4x^2$ _____

7. $\left(\frac{3}{5}\right)^3 = $ _____ $\frac{27}{125}$ _____

8. $24x^3 \div -2x^5 = $ _____ $\frac{-12}{x^2}$ _____

9. $3x^4 \times 5x^2 = $ _____ $15x^6$ _____

10. $(-5)^{-2} \times (-5)^5 = $ _____ $(-5)^3$, or -125 _____

11. $(4x)^0 \times (-2x)^2 = $ _____ $1 \times 4x^2$, or $4x^2$ _____

12. $9x^4 - 4x^3 = $ _____ $9x^4 - 4x^3$ _____

Algebra Readiness • Practice **183**

Lesson 4B

Chapter 22

Name _____ **Date** _____

Exponents and Applications: Lesson 4B

Practice

Each of the following simplifications has an error. Find the error. Explain what should have been done. Then simplify the expression correctly following the 5 Basic Rules. Assume all variables are not equal to zero.

1. $-4x^5 \times -4x^5 = -8x^5$
   need to multiply the numbers and add the exponents
   The correct simplification is _____ $-4x^5 \times -4x^5 = 16x^{10}$ _____

2. $14x^8 \div 7x^4 = 2x^2$
   need to subtract the exponents, not divide them
   The correct simplification is _____ $14x^8 \div 7x^4 = 2x^4$ _____

3. $x^{-6} \div x^{-2} = x^{12}$
   need to subtract the exponents, not multiply them
   The correct simplification is _____ $x^{-6} \div x^{-2} = x^{-4}$ _____

4. $(5x^2)^3 = 15x^6$
   must apply power of 3 to the 5 as well, not multiply 5 times 3
   The correct simplification is _____ $(5x^2)^3 = 125x^6$ _____

Simplify each expression using the 5 Basic Rules, PEMDAS, the distributive property, and the rules for signed numbers. Assume all variables do not equal 0.

5. $x^4 \times x^{-3} = $ _____ x^1, or x _____

6. $-3x^2(8x-1) = $ _____ $-24x^3 + 3x^2$ _____

7. $\left(\frac{2x}{7}\right)^2 = $ _____ $\frac{4x^2}{49}$ _____

8. $-6x^5 \div -3x^2 = $ _____ $2x^3$ _____

9. $-2x^2 \times 8x^2 = $ _____ $-16x^4$ _____

10. $(-4)^{-3} \times (-4)^2 = $ _____ $(-4)^{-1}$, or $-\frac{1}{4}$ _____

11. $(3x)^0 \times 2x = $ _____ $1 \times 2x$, or $2x$ _____

12. $-5x^3 - 4x^5 = $ _____ $-5x^3 - 4x^5$ _____

184 Algebra Readiness • Practice

---

244 **Algebra Readiness • Practice Answers**

Variables in Equations: Lesson 1A

Practice

Draw a line from the expression in Column 1 to its simplification in Column 2.

1. $5x^2 + 7x^2$ E   **A.** $30x^3$

2. $(-3x^2)(-5x^2)$ D   **B.** $-5x^4$

3. $6x^4 - 9x^4$ B   **C.** $-8x^6$

4. $7x^2 + 4x - 5 + 6x^2 - 8$ F   **D.** $15x^4$

5. $120x^5 \div 4x^2$ A   **E.** $12x^2$

6. $(3x^2 - 5x^2)^3$ C   **F.** $13x^2 + 4x - 13$

Simplify. Assume all variables do not equal 0.

7. $8x^2 - 11x^2 + 5x =$ __$-3x^2 + 5x$__

8. $-2x(x - 6) + 4x^2 =$ __$2x^2 + 12x$__

9. $-100x^6 \div 4x^4 =$ __$-25x^2$__

10. $\frac{1}{4}(8x^3 - 12x^2 - 20x) =$ __$-2x^3 - 3x^2 - 5x$__

11. $(-2x^2)(5x^2) + 8x^3 =$ __$-10x^4 + 8x^3$__

12. $3x(2x + 1) - 8x =$ __$6x^2 - 5x$__

Variables in Equations: Lesson 1B

Practice

Draw a line from the expression in Column 1 to its simplification in Column 2.

1. $9x^4 - 12x^4$ E   **A.** $-4x^6$

2. $40x^8 \div -10x^2$ A   **B.** $50x^6$

3. $(-5x^2)(-10x^4)$ B   **C.** $11x^3 - 14x$

4. $\frac{1}{2}(6x^3 - 4x^2 + 12x - 8)$ F   **D.** $3x^2$

5. $8x^3 - 5x + 3x^3 - 9x$ C   **E.** $-3x^4$

6. $(-3x)^2 + 6x^2$ D   **F.** $3x^3 - 2x^2 + 6x - 4$

Simplify. Assume all variables do not equal 0.

7. $3x - (-6x) - 4x^2 - 7x =$ __$2x - 4x^2$__

8. $(4x^3 - 6x^3)^2 =$ __$4x^6$__

9. $150x^6 \div -25x^2 =$ __$-6x^4$__

10. $12x^2 - 18x^3 \div 9x + 3 =$ __$10x^2 + 3$__

11. $8x + (-3x)(2x) - 10x =$ __$-2x - 6x^2$__

12. $15x(3x^2 - 5) =$ __$45x^3 - 75x$__

Variables in Equations: Lesson 2A

Practice

The following equations have been solved for you. After each step, write the property, process, or rule that was used to get to that form of the equation. Two of the more common processes are *combine like terms* and *simplify arithmetic*.

$$-4a - 12 = 3a + 9$$

1. $-4a - 12 - 3a = 3a + 9 - 3a$ __Addition property of equality__

2. $-7a - 12 = 9$ __Combine like terms__

3. $-7a - 12 + 12 = 9 + 12$ __Addition property of equality__

4. $-7a = 21$ __Combine like terms__

5. $\frac{-7a}{-7} = \frac{21}{-7}$ __Multiplication property of equality__

6. $a = -3$ __Simplify arithmetic__

Solve. Write each step, and be prepared to explain the property, rule, or process that you used.

7. $w - 4 = 10$ __$w = 14$__

8. $-6 = y + 5$ __$y = 11$__

9. $-21 = -3x$ __$x = 7$__

10. $\frac{2}{5}a = 12$ __$a = 30$__

11. $-3a - 1.4 = 5.2$ __$a = -2.2$__

12. $\frac{d}{5} = -10$ __$d = -50$__

13. $-4m + 6 = -14$ __$m = 5$__

14. $6c - 8c + 7 = -9$ __$c = 8$__

Variables in Equations: Lesson 2B

Practice

The following equations have been solved for you. After each step, write the property, process, or rule that was used to get to that form of the equation. Two of the more common processes are *combine like terms* and *simplify arithmetic*.

$$7m - 4m - 20 = -6 + 10$$

1. $3m - 20 = -6 + 10$ __Combine like terms__

2. $3m - 20 = 4$ __Combine like terms__

3. $3m - 20 + 20 = 4 + 20$ __Addition property of equality__

4. $3m = 24$ __Combine like terms__

5. $\frac{3m}{3} = \frac{24}{3}$ __Multiplication property of equality__

6. $m = 8$ __Simplify arithmetic__

Solve. Write each step, and be prepared to explain the property, rule, or process that you used.

7. $h - 9 = 10$ __$h = 19$__

8. $\frac{2}{3}y = -12$ __$y = -18$__

9. $\frac{m}{-3} = 12$ __$m = -36$__

10. $132 = 8y + 3y$ __$y = 12$__

11. $-5k + 2.6 = 1.1$ __$k = 0.3$__

12. $t + 3t - 5 = 11$ __$t = 4$__

13. $-8d + 6 = 3d - 5$ __$d = 1$__

14. $7n - 2n + 10 = 4n$ __$n = -10$__

Practice

There is an error in each simplification. Explain what the error is and simplify the expression correctly.

1. $-5(2a - 4) = -10a - 20$

The error is _____ sign error in the second multiplication _____

The correct simplification is _____ $-10a + 20$ _____

2. $(3x)^2 \times 2x = 3x^2 \times 2x = 6x^3$

The error is _____ did not take the 3 to the second power _____

The correct simplification is _____ $18x^3$ _____

There is an error in the solution for each equation. Explain what the error is and solve the equation correctly.

3.
$$-8m + 6m + 3m = 5 + 4$$
$$-8m + 9m = 5 + 4$$
$$-1m = 9$$
$$\frac{-1m}{-1} = \frac{9}{-1}$$
$$m = -9$$

The error is $-8m$ and $9m$ are
combined incorrectly

The correct solution is ___ $m = 9$ ___

4.
$$-4(2a + 6) - 10 = 6 - 8$$
$$-8a + 24 - 10 = 6 - 8$$
$$-8a + 14 = -2$$
$$-8a + 14 - 14 = -2 - 14$$
$$-8a = -16$$
$$\frac{-8a}{-8} = \frac{-16}{-8}$$
$$a = 2$$

The error is the sign on the second
multiplication is wrong

The correct solution is ___ $a = -4$ ___

Solve each equation. Show your steps.

5. $5y - 7y - 3y = 3^2 + 11$

$y = $ ___ -4 ___

6. $-3(2a - 1) + 6 = 3a$

$a = $ ___ 1 ___

7. $4(p - 3) + 2(3p - 1) = 5 + 1$

$p = $ ___ 2 ___

8. $\frac{2}{5}(10d - 15) + 3d = 3^3 - 5$

$d = $ ___ 4 ___

Practice

There is an error in each simplification. Explain what the error is and simplify the expression correctly.

1. $-4d(-3d + 4d^2) = -4d(d^2) = -4d^3$

The error is _____ combined unlike terms _____

The correct simplification is _____ $12d^2 - 16d^3$ _____

2. $-5a(6a^2 - 3a + 4) = -30a^3 + 15a - 20a = -30a^3 - 5a$

The error is _____ exponent error in second part of distributive _____

The correct simplification is _____ $-30a^3 + 15a^2 - 20a$ _____

There is an error in the solution for each equation. Explain what the error is and solve the equation correctly.

3.
$$3m - 7m + 2m - 4 = 12$$
$$-4m + 2m - 4 = 12$$
$$2m - 4 = 12$$
$$2m = 16$$
$$m = 8$$

The error is they combined the $-4m$
and $2m$ incorrectly

The correct solution is ___ $m = -8$ ___

4.
$$5(3a - 4) = 4^2$$
$$15a - 4 = 4^2$$
$$15a - 4 = 16$$
$$15a = 20$$
$$\frac{15a}{15} = \frac{20}{15}$$
$$a = \frac{4}{3}$$

The error is did not distribute the 5
to the 4

The correct solution is ___ $a = 2\frac{2}{5}$ or 2.4 ___

Solve each equation. Show your steps.

5. $6t - 4t + 3t - t = 3^2 - 1$

$t = $ ___ 2 ___

6. $-5(2c - 3) + 4c = 18$

$c = $ ___ $-\frac{1}{2}$ or -0.5 ___

7. $4(2n - 5) - 3(n + 2) = 8 - 9$

$n = $ ___ 5 ___

8. $2w - 7 = 6w - 5$

$w = $ ___ $\frac{-1}{2}$ ___

Practice

This equation has been solved and the solution has been proved. Write the explanation for what was done to reach each step.

Solution:

$-5(3t - 4) + 10 = 5^2$

		Explanation:
1.	$-15t + 20 + 10 = 5^2$	Distribute -5
2.	$-15t + 30 = 5^2$	Combine like terms
3.	$-15t + 30 = 25$	Simplify exponent
4.	$-15t + 30 - 30 = 25 - 30$	Addition property of equality
5.	$-15t = -5$	Combine like terms
6.	$\frac{-15t}{-15} = \frac{-5}{-15}$	Multiplication property of equality
7.	$t = \frac{1}{3}$	Simplify arithmetic

Proof:

$-5(3t - 4) + 10 = 5^2$

8.	$-5[3(\frac{1}{3}) - 4] + 10 = 5^2$	Substitute $\frac{1}{3}$ for t
9.	$-5[1 - 4] + 10 = 5^2$	Multiply 3 times $\frac{1}{3}$
10.	$-5(-3) + 10 = 5^2$	Simplify arithmetic in parentheses
11.	$15 + 10 = 5^2$	Multiply -5 times -3
12.	$25 = 5^2$	Combine numerical terms
13.	$25 = 25$	Simplify exponent

Solve each equation, and prove your answer. Show your work.

14. $4a - 6a + 8a = 5 \times 8 + 2$

$a = $ ___ 7 ___

15. $-3(2t - 5) + 2(t - 8) = 3$

$t = $ ___ -1 ___

Practice

This equation has been solved and the solution has been proved. Write the explanation for what was done to reach each step.

Solution:

$3(-2a + 4) + 4a - 8 = 3^2 + 1$

		Explanation:
1.	$-6a + 12 + 4a - 8 = 3^2 + 1$	Distribute 3
2.	$-2a + 4 = 3^2 + 1$	Combine terms
3.	$-2a + 4 = 9 + 1$	Simplify exponent
4.	$-2a + 4 = 10$	Combine like terms
5.	$-2a + 4 - 4 = 10 - 4$	Addition property of equality
6.	$-2a = 6$	Combine numerical terms
7.	$\frac{-2a}{-2} = \frac{-6}{-2}$	Multiplication property of equality
8.	$a = -3$	Simplify arithmetic

Proof:

$3(-2a + 4) + 4a - 8 = 3^2 + 1$

9.	$3[-2(-3) + 4] + 4(-3) - 8 = 3^2 + 1$	Substitute -3 for a
10.	$3[6 + 4] - 12 - 8 = 3^2 + 1$	Multiply
11.	$3(10) - 12 - 8 = 3^2 + 1$	Simplify arithmetic in parentheses
12.	$30 - 12 - 8 = 3^2 + 1$	Multiply 3 times 10
13.	$30 - 12 - 8 = 9 + 1$	Simplify exponent
14.	$10 = 10$	Combine numerical terms

Solve each equation, and prove your answer. Show your work.

15. $5y + 3y - 2y + 8 = -16$

$y = $ ___ -4 ___

16. $3(2m - 4) - 2(m + 1) = 3^2 + 1$

$m = $ ___ 6 ___

Equations and Inequalities: Lesson 1A

Practice

Translate each of these sentences into an equation. Solve the equation. To prove that your answer is correct, read the problem with the answer in place. If it makes a true statement, write *true*.

1. The difference between a number and 7 is 6. Find the number.

Equation: $n - 7 = 6$

Answer: number is 13

Proof: The difference between 13 and 7 is 6. True.

4. The quotient of a number and 6 is 12. Find the number.

Equation: $\frac{n}{6} = 12$

Answer: number is 72

Proof: The quotient of 72 and 6 is 12. True.

2. The product of −4 and a number is 28. Find the number.

Equation: $-4n = 28$

Answer: number is −7

Proof: The product of −4 and −7 is 28. True.

5. Five less than three times a number is 22. Find the number.

Equation: $3n - 5 = 22$

Answer: number is 9

Proof: Five less than 3 times 9 is 22. True.

3. The sum of −3 and a number is −8. Find the number.

Equation: $-3 + n = -8$

Answer: number is −5

Proof: The sum of −3 and −5 is −8. True.

6. Three times the difference of a number and 5 is 18. Find the number.

Equation: $3(n - 5) = 18$

Answer: number is 11

Proof: Three times the difference between 11 and 5 is 18. True.

Equations and Inequalities: Lesson 1B

Practice

Translate each of these sentences into an equation. Solve the equation. To prove that your answer is correct, read the problem with the answer in place. If it makes a true statement, write *true*.

1. The difference between a number and 2 is 8. Find the number.

Equation: $n - 2 = 8$

Answer: number is 10

Proof: The difference between 10 and 2 is 8. True.

4. The quotient of a number and −4 is 20. Find the number.

Equation: $\frac{-n}{4} = 20$

Answer: number is −80

Proof: The quotient of −80 and −4 is 20. True.

2. The product of −7 and a number is −56. Find the number.

Equation: $-7n = -56$

Answer: number is 8

Proof: The product of −7 and 8 is −56. True.

5. Six more than twice a number is −12. Find the number.

Equation: $2n + 6 = -12$

Answer: number is −9

Proof: Six more than two times −9 is −12. True.

3. The sum of −5 and a number is −10. Find the number.

Equation: $-5 + n = -10$

Answer: number is −5

Proof: The sum of −5 and −5 is −10. True.

6. Negative four times the sum of a number and 3 is 16. Find the number.

Equation: $-4(n + 3) = 16$

Answer: number is −7

Proof: −4 times the sum of −7 and 3 is 16. True.

Equations and Inequalities: Lesson 2A

Practice

This inequality has been solved. For each step, fill in the blank with the explanation for the process that was used to reach that step.

$4x - 8 - x + 3 \geq 16$

1. $3x - 5 \geq 16$ — Combine terms

2. $3x - 5 + 5 \geq 16 + 5$ — Add 5 to both sides

3. $3x \geq 21$ — Combine number terms

4. $\frac{3x}{3} \geq \frac{21}{3}$ — Divide both sides by 3

5. $x \geq 7$ — Simplify arithmetic

6. (number line 0–10) — Filled circle because ≥

Solve each inequality and graph the solution. Watch out for negative coefficients.

7. $2x - 3 < 11$

Solution is $x \geq 7$

8. $-4(x + 3) + 5 > 1$

Solution is $x < -2$

9. $3x - 5 \geq -26$

Solution is $x \geq -7$

10. $2(3x + 1) - 5(x - 2) \leq 9$

Solution is $x \leq -3$

Equations and Inequalities: Lesson 2B

Practice

This inequality has been solved. For each step, fill in the blank with the explanation for the process that was used to reach that step.

$6(2x + 3) - 8x < 20$

1. $12x + 18 - 8x < 20$ — Distribute 6

2. $4x + 18 < 20$ — Combine variable terms

3. $4x + 18 - 18 < 20 - 18$ — Subtract 18 from both sides

4. $4x < 2$ — Combine number terms

5. $\frac{4x}{4} < \frac{2}{4}$ — Divide both sides by 4

6. $x < \frac{1}{2}$ — Simplify arithmetic

7. (number line −10 to 10) — Open circle because <

Solve each inequality and graph the solution. Watch out for negative coefficients.

8. $5x - 2 > -27$

Solution is $x > -5$

10. $4x + 6 - 8x - 2 < 12$

Solution is $x > -2$

9. $-3(2x + 1) + 8 \leq -13$

Solution is $x \geq 3$

Algebra Readiness • Practice Answers   247

Chapter 24

Name _____ Date _____

Equations and Inequalities: Lesson 3A

Practice

Write and solve each word problem using an inequality, and then answer the question.

1. Meyer's Recycling Service has two pricing structures for its customers. Every container that is put out for pick up by the service must have a "paid" sticker on it. The fees for this service are outlined below:

   Plan A: $8 per month plus $0.25 per sticker

   Plan B: $5 per month plus $0.75 per sticker

   How many containers would a customer need to put out in a month's time for Plan A to be more cost effective than plan B?

Fill in the following table and then answer the question.

Steps	Problem
1) Identify the variable	x = number of stickers
2) Express all unknown quantities in terms of the variable	Plan A = $8 + 0.25x$ Plan B = $5 + 0.75x$
3) Set up the model	Plan A cost < Plan B cost
4) Solve and conclude	$8 + 0.25x < 5 + 0.75x$ $3 + 0.25x < 0.75x$ $3 < 0.50x$ $6 < x$ If a customer puts out more than 6 containers per month, Plan A costs less than Plan B.

Write and solve each word problem using an inequality.

2. Four times the sum of a number and 5 is less than 48. Find the numbers that will satisfy this relationship.

   Inequality: $4(n + 5) < 48$

   Solution: $n < 7$

Chapter 24

Name _____ Date _____

Equations and Inequalities: Lesson 3B

Practice

Write and solve each word problem using an inequality, and then answer the question.

1. Valley Electric Co–Op has two pricing structures for electric service for its rural customers. Each plan has a flat fee and then a rate for each kilowatt hour used.

   Plan A: $30 per month plus $0.04 per KWH (kilowatt hour)

   Plan B: $12 per month plus $0.06 per KWH

   How many kilowatt hours would a customer need to use in a month's time for Plan A to be more cost effective than plan B?

Fill in the following table and then answer the question.

Steps	Problem
1) Identify the variable	x = number of KWH
2) Express all unknown quantities in terms of the variable	Plan A = $30 + 0.04x$ Plan B = $12 + 0.06x$
3) Set up the model	Plan A cost < Plan B cost
4) Solve and conclude	$30 + 0.25x < 12 + 0.06x$ $18 + 0.04x < 0.06x$ $18 < 0.02x$ $900 < x$ If a customer uses more than 900 KWH per month, Plan A costs less than Plan B.

Write and solve each word problem using an inequality.

2. The product of 6 and a number, decreased by 9, is less than or equal to 15. Find the numbers that will satisfy this relationship.

   Inequality: $6n - 9 \leq 15$

   Solution: $n \leq 4$

Chapter 24

Name _____ Date _____

Equations and Inequalities: Lesson 4A

Practice

Give an example of each of the following mathematical terms.
Answers will vary.

1. Exponent:
   _____

2. Variable:
   _____

3. Multiplication Property of Equality:

4. Like terms:
   _____

5. Distributive Property:
   _____

6. Reciprocal:

Solve the following word problem: Three times the sum of a number and 4 is equal to the difference between that number and 10. Find the number.

7. Equation: $3(n + 4) = n - 10$

8. Solution: $n = -11$

9. Proof: Three times the sum of −11 and 4 is equal to the difference between −11 and 10.

Solve the inequality $-3(x + 2) + 7x \leq 2x - 10$. Show your work and write the explanation for what was done to reach each step. Graph your solution on the number line.

10. Solution: $x \leq -2$

   Graph: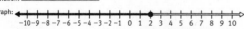
   -10 -9 -8 -7 -6 -5 -4 -3 -2 -1 0 1 2 3 4 5 6 7 8 9 10

Chapter 24

Name _____ Date _____

Equations and Inequalities: Lesson 4B

Practice

Give an example of each of the following mathematical terms.
Answers will vary.

1. Opposite:
   _____

2. Equation:
   _____

3. Addition Property of Equality:

4. Unlike terms:
   _____

5. Associative Property:
   _____

6. Base:

Solve the following word problem: Four times the difference between a number and 7 is equal to the sum of that number and 5. Find the number.

7. Equation: $4(n - 7) = n + 5$

8. Solution: $n = 11$

9. Proof: Four times the difference between 11 and 7 is equal to the sum of 11 and 5.

Solve the inequality $-2(x - 5) - 4 > -4x + 18$. Show your work and write the explanation for what was done to reach each step. Graph your solution on the number line.

10. Solution: $x > 6$

   Graph:
   -10 -9 -8 -7 -6 -5 -4 -3 -2 -1 0 1 2 3 4 5 6 7 8 9 10
